WHERE MEDICINE WENT WRONG
Rediscovering the Path to Complexity

STUDIES OF NONLINEAR PHENOMENA IN LIFE SCIENCE

Editor-in-Charge: Bruce J. West

Vol. 1   Fractal Physiology and Chaos in Medicine
by B J West

Vol. 2   Patterns, Information and Chaos in Neuronal Systems
edited by B J West

Vol. 3   The Lure of Modern Science — Fractal Thinking
by B J West & B Deering

Vol. 4   Physical Theory in Biology — Foundations and Explorations
edited by C J Lumsden, W A Brandts & L E H Trainor

Vol. 5   Nonlinear Dynamics in Human Behavior
edited by W Sulis & A Combs

Vol. 6   The Complex Matters of the Mind
edited by F Orsucci

Vol. 7   Physiology, Promiscuity, and Prophecy at the Millennium: A Tale of Tails
by B J West

Vol. 8   Dynamics, Synergetics, Autonomous Agents: Nonlinear Systems Approaches to Cognitive Psychology and Cognitive Science
edited by W Tschacher & J-P Dauwalder

Vol. 9   Changing Mind: Transitions in Natural and Artificial Environments
by F F Orsucci

Vol. 10   The Dynamical Systems Approach to Cognition: Concepts and Empirical Paradigms based on Self-Organization, Embodiment, and Coordination Dynamics
edited by W Tschacher & J-P Dauwalder

Vol. 11   Where Medicine Went Wrong: Rediscovering the Path to Complexity
by B J West

Studies of Nonlinear Phenomena in Life Science – Vol. 11

WHERE MEDICINE WENT WRONG

Rediscovering the Path to Complexity

Bruce J. West

US Army Research Office, USA

World Scientific

NEW JERSEY · LONDON · SINGAPORE · BEIJING · SHANGHAI · HONG KONG · TAIPEI · CHENNAI

Published by

World Scientific Publishing Co. Pte. Ltd.
5 Toh Tuck Link, Singapore 596224
USA office: 27 Warren Street, Suite 401-402, Hackensack, NJ 07601
UK office: 57 Shelton Street, Covent Garden, London WC2H 9HE

British Library Cataloguing-in-Publication Data
A catalogue record for this book is available from the British Library.

The image on the front cover by courtesy of Sharon S. West.

WHERE MEDICINE WENT WRONG
Rediscovering the Path to Complexity
Studies of Nonlinear Phenomena in Life Science — Vol. 11

Copyright © 2006 by World Scientific Publishing Co. Pte. Ltd.

All rights reserved. This book, or parts thereof, may not be reproduced in any form or by any means, electronic or mechanical, including photocopying, recording or any information storage and retrieval system now known or to be invented, without written permission from the Publisher.

For photocopying of material in this volume, please pay a copying fee through the Copyright Clearance Center, Inc., 222 Rosewood Drive, Danvers, MA 01923, USA. In this case permission to photocopy is not required from the publisher.

ISBN 981-256-883-2 (pbk)

Typeset by Stallion Press
Email: enquiries@stallionpress.com

Printed by FuIsland Offset Printing (S) Pte Ltd, Singapore

Contents

Prologue		ix
Acknowledgments		xiii
1. Chance and Variation		**1**
1.1.	The myth of equality	3
1.2.	Slaughter's Café	10
1.3.	What are the odds?	14
1.4.	Odds against smallpox	22
1.5.	Information and chance	33
2. The Expectation of Health		**37**
2.1.	Control and cybernetics	40
2.2.	Temperature regulation	46
2.3.	Respiration regulation	50
2.4.	Cardiac regulation	57
2.5.	Averages are not sufficient	65
3. Even Uncertainty has Laws		**69**
3.1.	Randomness and measurement	72
3.2.	Chaos and determinism	87
3.3.	Wisdom is not static	94
3.4.	A new tradition	111

4. The Uncertainty of Health	**119**
4.1. The different kinds of scientists	124
4.2. The Emperor in exile	130
4.3. What is wrong with the law of errors	138
4.4. The inverse power-law distribution	144
4.5. How the physical and life sciences are different	156
4.6. Only the few matter	168
5. Fractal Physiology	**173**
5.1. Scaling in physiological data	176
5.2. Allometric relationships	182
5.3. Fractal heartbeats	191
5.4. Intermittent chaos and colored noise	198
5.5. Fractal breathing	204
5.6. Fractal gait	216
5.7. Fractal temperature	223
5.8. Fractal gut	228
5.9. Fractal neurons	231
5.10. Internetwork interactions	238
6. Complexity	**241**
6.1. Random networks	244
6.2. Scale-free networks	255
6.3. Controlling complexity	259
6.4. Allometric control	265
6.5. Disease as loss of control	273
7. Disease as Loss of Complexity	**283**
7.1. Pathological periodicities	285
7.2. Heart failure and fractal loss	289
7.3. Breakdown of gait	302
7.4. Summing up	310

Epilogue	**315**
References	**317**
Index	**327**

Prologue

It seems that the end of a century has historically motivated certain scientists to take stock of the progress that science has made over the preceeding 10 decades. Such soul searching usually begins with the proclamation that science has finished its job and solved all the fundamental problems concerned with understanding nature. At the end of the 19th century, one of the leaders of physics, Lord Kelvin, cautioned his students against going into science because all the important work had already been done and only engineering questions remained to be answered. There were only a few details to be clarified, such as understanding black body radiation and some paradox regarding the lumeniferous aether. At the turn of the 20th century, the solution to the problem of black body radiation led to quantum mechanics and the resolution of the contradictions regarding the aether gave the world special relativity. So no matter how brillant the person, predictions, in the absence of theory, should be left in the hands of the fortune tellers, shamans and charlatans.

At the end of the 20th century we also had those scientists that predicted the end of science. But lack of imagination does not constitute a proof. Lord Kelvin could not foresee quantum mechanics and therefore he concluded that the singularity resulting from the radiation from a black body was only a detail and nothing of substance remained to be done in physics. As we said, he could not have been more wrong. But unlike the end of the 19th century, the end of the 20th century did not identify a number of problems whose resolution would neatly wrap things up. The big questions that remained at the turn of the 21st century focused on nonlinear science, nonlinear networks and how to properly characterize complexity. These latter questions are open-ended and change with the context, which is to say, complexity forms barriers in physics, biology and medicine, but in very different ways.

Our interest in this book is in medicine, but that interest does not blind us to the fact that the human body is subject to all the laws of the physical world, that we are alive and therefore have a biological origin, and the body is a large chemical network. If we attempt to be overly specific about how the body functions we quickly become bogged down in an interpenetrating system of mutually dependent networks. The human body is certainly a system-of-systems, with the sympathetic and parasympathetic nervous systems, endrocrine system, respiratory system, gastrointestinal system and on and on. The discipline of medicine is the empirical study of the interactions among these various systems, as well as the interactions of the component parts within any specific system. Moreover, in its search for understanding, medicine seeks to repair these systems when they are damaged or diseased.

Our thesis in this book is that medicine, somewhere in its development, took a wrong turn. Furthermore, that wrong turn was closely tied to the fact that the human body is an incredibly complex structure and we human beings shy away from the complex and embrace the simple. Part of the attraction for simplicity has to do with the fact that we can understand simple things; complex things confuse and frustrate us. Therefore when confronted with a new phenomena we first try to simplify it, break it down into its component parts and see if we understand them. Like giving a complicated toy to a curious child who takes it apart in an effort to see how it works. In the 1950s, every teenage boy could take an internal combustion engine apart and put it back together, explaining how each of the parts contribute to propelling the car from zero to sixty in under ten seconds. But cars are, in fact, simple physical/chemical objects and we would attempt to understand the human body in the same way we understand the thermodynamics of the automobile. But alas we cannot. It is too complex.

So what do we do when we cannot implement this reductionist reasoning of pulling things apart in order to understand how they work? One thing that we do is to look for measures that we can use to determine how well the phenomenon we are examining functions. A few decades ago car engineers believed that the consumer wanted to be able to monitor the vital signs of a car so the dashboard was a collage of gauges indicating oil pressure, the revolutions per minute of the engine, engine temperature, whether the battery is taking or losing a charge, and so on. Of course, in modern cars most of these gauges have been replaced with lights that only go on when the given component of the car is in trouble, such as the oil light coming on just before the engine ceases up. The modern view is that the consumer is no longer able to interpret the complex of indicators, since s/he probably does not understand how the automobile works, and therefore the only information of value to the driver is when to bring the car to the specialist in the garage.

We have acknowledged that the car is a simple physical/chemical system, consequently the measures of how well the car is operating are correspondingly simple and useful. We have also applied the same kind of reasoning to the human body, providing the consumer with a few measures of how well the body is functioning in spite of its admitted complexity. In this way we monitor health through our heart rate, blood pressure, body temperature, breathing rate, gait and so on. Like the well-monitored car of the last century, we keep track of a wide array of subsystems within the body's system-of-systems. The fallacy is that the human body is not simple and these historical measures of health are not the quantities at which we should be looking. This book is about what replaces these traditional measures of health and why traditional physiology ought to be replaced with fractal physiology.

Acknowledgments

My heartfelt thanks go out to my friends and colleagues who have worked with me over the years in this most productive field of research. These include my good friends Mirek Latka, with whom I now collaborate over the Internet on homeodynamics, among other things; Nicola Scafetta, with whom I have developed some exceptionally useful mathematical models; and Richard Moon who has reaffirmed the caution one must take in developing biomedical models. In addition, I greatly appreciate and thank those who have given freely of their time in reading earlier versions of the manuscript: Robert Lindberg and Richard Moon for their detailed comments on style, comprehension and medical accuracy; Alan Mutch and Timothy Buchman for their comments on the overall presentation of the work; Ary Goldberger, for our early collaborations that set the direction for what came later, and finally the late Jonas Salk for sharing with me his philosophical perspective on medicine. In any event, I take full responsibility for any and all typos, misstatements and outright errors that remain. Finally, what I have learnt from my family, Sharon, Jason, Adrianna and Alex, Damien and Jennifer, has made doing this work a joy.

Chapter 1

Chance and Variation

This book is a tour through the technical developments that lead to the modern theory of medicine, or at least one version of such a theory. Being on a tour I do not spend much time in any one place, or give too much detail on technical issues, but attempt to paint in broad strokes, laying out what I consider to be the important background. I do, however, consider some concepts to be central to the discussion, concepts that are so obvious that they are rarely, if ever, questioned. Such concepts are like the water in which fish swim; the fish does not see the water because that is the very basis of its existence. The examination of such concepts in physiology, by a relatively small number of scientists, has produced a revolutionary view of medicine over the past two decades that has gained more than a foothold in medical research. This new view has been collected under the rubiric of *fractal physiology*. The application of nonlinear dynamics and fractal statistics to physiologic phenomena has enabled physicians to uncover and interpret a new richness in physiologic time series, but I am getting ahead of my story.

It is generally prudent for me to consult a map before starting out on a journey, particularly if I am traveling to somewhere I have not been before. But even if I have made the trip previously, I find that it does not hurt to refresh my memory about unmarked turns and possible areas of congestion. For this reason I begin each chapter with an intellectual map of what is to be discussed, because many of the places I visit, if not new to the reader, will be approached from a unique perspective. Thus, there are many readers who have driven into New York City, but few have parachuted into Times Square from an airplane. My experience of the city would be quite different in the two cases and the conclusions I might draw regarding what is, or is not, important about the city would also be very different.

In this first chapter I argue that the history of western civilization has formed the way in which we, its progeny, view the world. This mundane observation, made in every history class on every college campus, often fails to make the case for specific ideas. The big concepts like democracy, capitalism, socialism and the like, are more than adequately discussed, even if I might take issue with the relative emphasis. In the present analysis I am more concerned with a handful of little ideas rather than big ones. Little ideas like whether the world is a fundamentally fair place and the origin of the idea of fairness. It will surprise many that fairness developed at the intersection of law and gambling. From this nexus of ideas, other notions took shape, such as ambiguity, averages and eventually certainty about the future. These are the things with which I will be concerned, because the little ideas, like equality and coin tossing, have done more to influence how most individuals within our society think about the world than have the big ideas.

Equality is a well defined mathematical concept, but it loses precision when applied in a social context, since no two people are exactly alike. To trace the myth of social equality requires that I develop some elementary statistical concepts and examine how these arcane ideas crept into the dicussion of social well being. In particular I am interested in how medicine was affected by such pubic health decisions as inoculating a population against smallpox in the 17th century. This background is useful in order to appreciate how a predominantly agrarian society molded some of the less obvious concepts supporting our present day view of medicine. Furthermore, I argue that by rejecting these fundamental assumptions many of today's questions concerning certain diseases can be answered, at least in part.

I outline how to reason about luck and chance, because it was through such reasoning that society determined to adopt average measures as the way to think about complex systems like the human body. These arguments can become very convoluted and formal, so I also introduce the idea of information as a simpler way of characterizing complex phenomena. The discussion regarding information and even probability is elementary and will be taken up repeatedly in subsequent chapters as the need arises, but we avoid introducing equations and attempt to communicate the content without the formalism. The emphasis is on ideas.

1.1. The myth of equality

Today society has the telephone, the Internet, the FAX machine, email, and any of a number of other technologies by which its members can remain in contact with one another. It is not an exaggeration to say that communication in the 21st century is instantaneous throughout most of the industrialized world. Like communication, when I am faced with a problem, my response for obtaining a solution tends to be either immediate or never. Modern day problem solving methods often reject solutions that have intermedicate time scales, forcing those solutions that can be immediately executed over those that require maturation times for their implementation. It is almost impossible to trace today's archetypes back to their non-industrialized beginnings, but that is what I need to do, at least in part, in order to clarify the limitations of understanding in modern medicine due to complexity.

In the 16th century 90% of Europe's population lived on farms or in small towns and their view of the world was very different from what ours is today. This difference is due, in part, to how people interacted with the physical world. The foundation of an agrarian society was land, animals, people and tools, all motivated by the creation and storage of food. The typical latrine was a small wooden structure tens of meters from the house, with one, two or three holes, each the size of a pie plate, cut into a board spanning a ditch half filled with biowaste, with no partitions to separate the often multiple simultaneous users of the facilities. Water was obtained from a well, using either a wynch to lower and raise a bucket at the end of a rope, or using a pump that invariably needed to be primed. The bread on the table was baked in the farmer's kitchen in a wood burning stove, using flour that was locally milled. The meat was butchered in the fall and preserved by smoking, salting or canning in grease for the long winter. Fresh fruit and vegetables were determined by what could be picked from the trees or pulled from the farmer's garden and stored in a root cellar.

One example of a relatively stable agrarian society was the Hälsingland farms in Sweden[1]:

> A Hälsingland farm is a unit of production, and for it to function there had to be a balance between land and animals, people and tools, and buildings. If something is missing, the equilibrium is disturbed and the whole system falls apart.

These farms actually began around the beginings of the Iron Age and continued to grow throughout the Middle Ages, retaining a constant level in the 250-year period from 1500 to 1750. In such a successful society one would expect the population to grow, which it did, but the labor excess did not become farmers, but rather they became fishermen, tradesmen and craftsmen. The migration from the land was an adaptation made by the people to the constraints of the society into which they were born, such as the amount of land available for farming.

However, Europe was not homogeneous, and the ideal of the Hälsingland farms was only partially realized elsewhere. Regardless of the degree of success of the social system, the farm and its demands moulded the world view of the rural and urban populations alike. A farmer would get up before sunrise to work the farm: milk cows, gather eggs, feed the stock; and then s/he would come in for breakfast. Whenever possible the morning meal would be large, consisting of meat, usually pork or beef, fried in a heavy grease, and eggs, bread lathered with cream rich butter and containers of milk. Potatoes would later be added to the farmer's diet, often replacing the more expensive and harder to obtain meat. Breakfast would provide the energy required for the day's work in the field. The time of year determined whether one plowed, planted or harvested — but it was always physical and occupied most of the farmer's time. When it was too hot to work in the field, often after the mid-day meal, there were always machines to be fixed, buildings to be repaired and animals to be tended. A farmer's work was never done.

The social interactions during the week were with one's family. The children always had to be taught, by both the mother and the father. Farming was hard manual work for both sexes, and contrary to common belief, the work was an intellectual challenge to those who were successful at it. Filling a barn with hay and a silo with grain, so the farm animals could eat through the winter, was not a simple task, recalling that only animals were available to carry or pull loads, and only the farmer's imagination determined the machines that could be used to leverage that horse power or human power. Farmers understood levers and simple machines, they also knew the work that running water can do, such as turning a paddle wheel to mill flour, and the wind driving a windmill blade to pump water from deep underground. These things had to be taught to one's children at the earliest age possible, because life was brief and uncertain, and one had to be able to make one's way in the world.

Walking along a country road every few minutes in Europe and somewhat longer in the colonies, a person encountered another farm. These neighbors were the second circle of socialization known to farmers and were the people one met in church on Sunday and in town on Saturday, when one went shopping for staples. These neighbors shared a common perspective on the weather, the price of crops, the marrying ages of various children and the successes and failures of both last year's and next year's harvest.

The individuals that survived childhood were generally strong, independently minded and confident in their ability to control the small world around them. In the 20 year period 1603–1624 in the city of London, 36% of all deaths occurred before the age of six. In Paris in the middle of the 18th century, approximately 20% of all children were left in Foundling Hospitals and of those dying, 30% were in such hospitals. This practice of abandoning children continued well into the 19th century (see Montroll and Badger[2] for a more complete discussion).

For many, the farm became too small a universe and they would strike out for the nearest big city, whether it was London, Paris, Rome or New York. The city was a dramatically different place from the 17th century farm. The farmer entered the city with a wagon of fruits and vegatables to sell, and the buyers were those that cooked for the city dwellers. This was the third and final level of socialization for the farmer. They talked, joked and argued with the wives of the merchants and artisans, servants from the wealthy families, and cooks from restaurants and taverns. All these buyers would compete with one another for the best and cheapest staples. The farmers were at the bottom of the food chain necessary to sustain the life of the city.

At this point in history the city was, as it is today, a more complex interactive structure than was the farm. The cities, by and large, grew without plan or design to satisfy the needs and inclinations of its occupants or those of the city fathers. No one knows what the precise mix of food delivered to the city and food eaten by the various strata of the city was; what the proper balance in the number of restaurants versus the number of garbage collectors was; how many hotels, hostels and private homes there were; what the number and kind of streets there were, what the various modes of transportation were, or what the size of the merchant class was. It was understood that the city needed various things to survive: food, housing, power, and water.

Individuals identified what things were necessary and how they could profit by supplying them to the city for a price. Those seeking their own profit performed a function that insured the survival of the city. The city in turn combined the contributions of the many and provided an environment that no one person could have realized on his/her own. This environment attracted other individuals that could provide additional goods and services needed by the expanding city.

The time interval between strangers in the city was fractions of second, rather than the minutes that separated even good friends in the country. Letters could be delivered by courier on the same day they were written, for those that could read. Newspapers documented rumors, developments, politics and the price of cotton, so that individuals could receive information on a daily rather than weekly basis. In the city the son no longer learned from the father, or the daughter from the mother; instead, the child was apprenticed at the age of eight or so to a master, to learn a trade or craft, and for most people this craft determined their life's path. Few individuals, once apprenticed, had the strength of purpose to abandon their apprenticeship, and go into a future in which their means of earning a living was not secure. In one sense the city replaced the family and this replacement has become almost complete with the blending of the 19th century industrial revolution with the 20th century information revolution.

The city and the farm made each other possible. The farm supplied the excess food necessary to feed the city, and the city provided opportunities for the empolyment of the extra labor generated by the large families on the farm. Montroll and Badger[2] argue that the potato, which was extremely nutritious, could be grown on poor land and had a yield per acre in nutritional value that was two to four times higher than grain; this then contributed to earlier marriages and higher birth rates. They go on to say that an acre of poor soil planted with potatoes could support a family of between six or eight. In part this explains the continuing population explosion that occurred in Europe, but which did not disrupt the more traditional Hälsingland farms of Sweden.

The excess labor from the farms, in each generation, that could not adapt to the system by becoming tradesmen or craftsmen, went to the city in search of fortune and adventure. The city was the market for the farmer's excess produce and it was also the ladder for each generation's ambition. The farmers were thought to be uncouth, ignorant and superstitious by the city folk, who in their turn were perceived as dandies, cynics and ungodly by their country cousins. Each was seen through the lens of what the viewer could do best, and was rarely acknowledged for their complementary abilities.

It may seem a bit odd, but it was out of this very strange looking world that the contemporary view of being human developed. In large part, it was the conflict between the remnants of the fuedal system and the notion of ownership in the post-Renaissance period that forced the examination of what it meant to be human. Philosophers struggled with the meaning of life, man's place in society, what property meant, whether slavery was acceptable and what constituted fairness. It is interesting that the concept of fairness came out of the synthesis of two very different arenas: law and games of chance.

In the 17th and 18th centuries, law was a very different social phenomenon from what it is today. The way law is accepted by the members of an agraian society is not how it is experienced by those in an urban society. On the farm, law, any law, was seen to be the imposition of the will of the strong on the weak; the domination of the minority by the majority; the disregard of the wants and needs of the individual by the caprice of the group. This perspective on the law was perhaps a consequence of the farmer's ability to stand alone, outside society for the most part, and support his family. The farmer's subsociety was sufficient to insure survival, pretty much independently of what happened in the larger society, probably with the exception of war. Since the farmer accepted little from society, s/he was unwilling to give up anything substantial to society. In the city the view of law was closer to today's perception, since a city, any city, could not survive without codes of conduct and methods of punishment. In the city laws came to be seen as protecting the poor and weak; the social contract that enabled the individual to profit from the abilities of the group; the protection of property ownership from revolution or government usurpage and finally the preservation of the state and therefore the city. Laws produced stability.

One of the fundamental concepts that developed out of law in this period was the notion of fairness. From one perspective, law is man's attempt to impose order on the interactions among individuals and thus make social discourse rational. In England, in part due to the backlash against the church and the aristocracy, the law assumed symmetry between contesting parties, that is, absent any evidence to the contrary two individuals in dispute were viewed as equivalent. It developed that no person's position under the law was seen as different from any other person's, until evidence to the contrary was introduced. Equivalence was an integral part of the notion of fairness, interchangeability of individuals under the law. Fairness is why everyone is innocent until proven guilty; it would be unfair to treat certain people differently from the majority. Thus, the very characteristics that made individuals distinguishable from one another was considered to be irrelevant to the law. So fairness became a doctrine of equivalence and required that society determine what the essential properties of being human are and to disregard those characteristics that make individuals different from one another.

During the same period a gentleman had three major diversions; drinking, wenching and gambling. The first two do not concern the present discussion, so I focus my attention on games of chance. At this time in history most people were illiterate and understanding arithmetic was somewhat of a black art. A person who could calculate the odds for a game of dice or devise a strategy for winning at *Whist* was considered in an almost mystical light. Consequently those "scientists" who had the misfortune to be born poor often made a living by casting horoscopes or calculating the odds for gentlemen consumed with the most recent gambling fashion.

1.2. Slaughter's Café

It was a cold rainy morning on London's east side when a small man entered Slaughter's Café in St. Martin's Lane and walked over to his usual table in the corner where he was able to catch the morning light coming through the window. The café had just opened and the smells of newly delivered baked bread and thick fresh coffee hung low around the table. The man placed a satchel containing ledgers, paper, pen and ink on the table and slowly seated himself anticipating his cup of morning coffee before his customers began to arrive. His rooms were cold and damp, and the dampness had gotten into his bones this morning causing him to shiver as he looked around the large room. He withdrew a sheath of papers from the satchel and began reading the last few pages of the closely spaced lettering of the manuscript. The shivering subsided as he read and the proprietor, knowing the loner's habits, brought him his coffee without invitation. The man barely acknowledged the owner's passage. This was his favorite time of day. He would have an hour or so to himself in which to think and write. He would recall the discussions with the great man from the previous evening and sift through their conversation to find what might have lasting value for his work. It was odd, the café standing empty of customers except for the small man hunched over his manuscript with a sly smile of contentment on his face. It was the late 17th century.

The rain stopped as it always did and the sun came out to dry the London streets. Slaughter's Café was now filled with smoke, the clatter of dishes and the unceasing din of conversation. The small man still sat at his table, but now he was one of many. The manuscript had been replaced by an odd assortment of scraps of paper and a rather well dressed man sat at his side earnestly talking about a game of dice and the possible outcomes one might expect.

"I was told that you could tell me how to win at throwing the dice," said the gentleman, not looking directly at the mathematician.

"Then you have been misinformed," responded the smaller man. "What I can tell you is how often a number comes up on the dice and give you the odds on which to bet. This will give you the best chance of winning, but it is not a guarantee."

The small man looked knowingly at his client. The question was one that he often heard from the gamblers that frequented his table. They were men of breeding who were not satisfied by eating, drinking and whoring, but also wanted the exhiliration of gambling. The hitch was, that almost to a man, they were ignorant of the fundamentals of mathematics. That was not a problem, however, because they could always buy the information they needed from men such as this Hugenoit refugee in Slaughter's Café. He was learned, and some said he was even brilliant, but he was poor and without the proper sorts of connections. Such men sell their talents quite reasonably, so the small man patiently explained.

"The number of different combinations one can make with two die is twelve, since there are six numbers on each dice. If we count the total number of ways we can make a seven, there are 4 and 3, 3 and 4, 2 and 5, 5 and 2, 6 and 1, 1 and 6, and no others. Therefore there are six ways in twelve to roll a seven. Therefore the odds against the number being a seven on any roll of the die is two to one. Understand?" The small man spoke slowly and clearly looking into the larger man's eyes the full time. In the realm of the mind he was far richer than his well dressed companion.

The gentleman looked somewhat impatiently at the mathematician and said, "My good man, I do not need a lesson on these things. I only want to know the final odds. So you calculated that the odds are two to one against my rolling a seven."

"Yes, that is correct," said the mathematician.

"Good. Now what are the odds against a five coming up on a roll?"

The mathematician nodded and slowly began to explain how one computes the odds for making a five. He knew his client would not understand, and might even become annoyed, but more importantly he knew from experience that if he just gave him the results without explanation, the client would never return. If it looked too easy to calculate the odds, then it could not be worth even the small amount he was paid. Thus, the mathematician did not heed the protests of the young gentleman, and subjected him to explanations, which he would never understand, so he would appreciate that this quiet, almost sulky, man, that calculated odds for a living at this corner table in a café, knew something valuable and quite mysterious. It was the mystery that drew the customers back.

As the day wore on, a different kind of client could be seen talking with our mathematician. Men, dressed not in gentlemen's finery, but in the solid broadcloth of merchants and tradesmen.

"We have ships embarking for the Americas with a full compliment of colonists. They have each paid for their passage and we expect the ships to return with a full load of tabacco from Virginia. In the past year two ships have been lost and a total of seven have completed the journey. Considering the cost of each ship, the potential loss of the ship being lost on the way over, or lost on the way back, can you determine for how much the owners should insure each ship? What should the insurance premium be?"

The questions asked by the insurance men were more interesting than those asked by the gamblers, at least, in part, because some aspects of the problems were always a little different. He did not need to explain in detail how the calculations were made, because the insurance man knew arithmetic and also, it was cheaper to pay the mathematician the few pennies he charged to solve the problem than it was for him to waste time doing it and possibly getting it wrong.

It had been a productive day. He had written four pages in his manuscript, and had collected half a crown by calculating odds for gamblers and insurance agents. Fortunately no one had come in to have their horoscope read. Although that generally paid more money than calculating odds, he never felt quite comfortable discussing the effect of the alignment of the planets on a person's life. Especially since he had started discussing these and other matters with the great man. It was approaching the dinner hour, when Sir Isaac Newton would be walking home from the Mint. As was his habit he would stop in at the café and bring the young mathematician to his home nearby. That was the highlight of de Moivre's day, when after dinner he would sit in Sir Isaac's den sipping port and discussing natural philosophy.[3]

Most scientists of the 17th and 18th centuries fell into two categories: those who were members of the aristocracy, and those who were not and, like artists, had benefactors. A third group is typified by Abraham de Moivré, a Huguenot refugee from France who took shelter in England. De Moivré was a brilliant mathematician who could not afford to become a student and was self-taught. He managed to survive by tutoring in mathematics, calculating annuities and devising new games of chance for which he could calculate the odds of winning. Many of the questions he answered for the gentry are contained in his 1718 book *The Doctrine of Chances* with the subtitle *A Method of Calculating the Probability of the Events of Play*. He was, moreover, a friend of Sir Isaac Newton, who would come by Slaugther's Cafe after a day's work as Master of the Mint.[3] De Moivré spent his days in the coffeehouse, determining the insurance on cargo being shipped to the new world and calculating odds on games of chance for gentlemen seeking an advantage over their friends. A mathematician could make a living devising new gambling games for the aristocracy, if not by teaching them the secrets of the discipline.

De Moivré stripped the games of chance down to their bare essentials and was able to calculate the odds for and against the occurence of any given outcome. Consider the simplest of wagers, that being on the toss of a coin. The bet is that a head will appear on the toss. There are two possible outcomes, a head or a tail. Heads you win, tails you lose. There are as many ways of winning as there are of losing, therefore a fair bet is even money. You can win (lose) as much money as your oppenent loses (wins). One can bet that a head or a tail occurs on each toss and one or the other occurs without bias. Thus, a head or a tail is equally likely on each toss of the coin. Of course, that is not the only available bet. One could wager on the outcome of two tosses of the coin, yielding two heads, two tails or one of each. Here again the likelihood of each outcome is the same. Thus, fairness and the amount of a wager became intertwined.

1.3. What are the odds?

The reasoning of de Moivré involves the way a mathematician formulates the solution to scientific problems. It is therefore useful to review the ideas behind some of the simple models he constructed, since these ideas eventually found their way into the 20th century scientist's view of the world in general and medicine in particular. These mathematical ideas have interesting social implications but I make every effort to avoid the usual accompanying mathematical formalism. This discussion is necessary in order to have a clear idea of what is meant by probability and fair play, used later in discussing medical phenomena. If you feel comfortable with these concepts, then by all means skip this subsection and go on to the next. If, however, your probability calculus is a little rusty, this subsection has the virtue of brevity.

The simplest game of chance involves the flipping of a coin. Heads I win, tails I lose. But more interesting combinations of outcomes can be used to formulate various wagers. To calculate the odds for the different bets one can make involving a sequence of the two possible outcomes of a coin toss, de Moivré formulated the branch of algebra known as combinatorics, to determine the number of ways heads and tails can occur in a given number of tosses. For example, how many ways can six heads and four tails occur in 10 tosses of a coin? Not an easy question to answer by direct counting, so de Moivré invented a condensed way of counting the outcomes of coin tossing, now known as the binomial theorem. But I will spare you the algebraic equations, especially when words will serve equally well and perhaps communicate even better.

Let me make this coin tossing business a little clearer. Suppose I toss a coin three times and ask for the number of different ways I can obtain two heads out of the three tosses. Figure 1.1 indicates the network of all the possible outcomes of tossing the coin three times; the network of my possible futures has each outcome specified by a different time line. The network shows the two choices on the first toss at the bottom of the figure, but only one of which can be realized, which is to say, either a head or a tail will definitely come up. There are four possible time lines generated by the first two tosses, corresponding to four possible paths that can be realized. There are eight possibilities on the first three tosses and so on. The number of ways only two heads can arise out of the three tosses shown in the figure, can be determined by the three paths from the bottom of the network to the top that have only two heads and that can be traversed without retracing any steps. These paths are indicated by the dark lines. Consider the left most dark path; first a head is obtained, then a second head, followed by a tail. The next dark path,

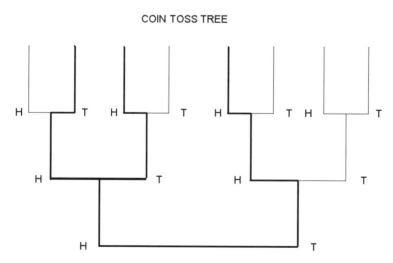

COIN TOSS TREE

Fig. 1.1: On the first toss, there are two possible outcomes a head denoted H or a tail denoted T. With the next toss, there are a total of four possible outcomes HH, HT, TH or TT and so on. Any given three tosses will trace one, and only one, possible path from the bottom to the top of the network. The dark lines indicate the paths that contain two heads only in three tosses.

moving from left to right, again starts from a head, is followed by a tail, then a second head. One can similarly trace out the remaining path. Starting from the bottom of the diagram, it is impossible to predict which of the possible time lines will actually materialize. In fact when I start out each of the futures is equally likely. However, at the end of the three tosses, one and only one path has been realized. In other words, all the possible futures collapse into a single reality, the one objectively experienced by the coin tosser.

Of course, the same is true of the number of paths for two tails out of three tosses. At any level, each of the outcomes is equally likely. Consider the three tosses in the example, each of the eight outcomes is equally likely. The fact that there is no information to discriminate among the eight outcomes before I start tossing a coin implies that each path, from the first to the third toss, is unique and a wager on any particular path would yield an equal payoff, that is, the same payoff for each path. Consider the path in which there is a HTH on the three tosses and compare this with the path in which there is a THH on the three tosses. There is only one way in the eight distinct outcomes that either sequence can occur, so one might reason that a fair wager would be one in which for each dollar I bet someone should be willing to put up seven, since there are seven ways another can win and only one way I can win. This direct proportionality between what I am willing to wager and how much I can win seems to be at the heart of fairness. This is the fundamental nature of fairness; you either see this or you do not. This one-to-one proportionality seems to be present even in primative societies and predates the abstractions of numbers and counting.

The betting becomes more subtle when the uniqueness of the paths are set aside and one bets on the total number of outcomes and not on the order in which the events occur. The outcome of two heads and one tail, when the order is not important, can be obtained in three different ways, that is, HHT, HTH and THH all become equivalent, having two heads and one tail, independent of the order in which they occur. When I do not care about the order, there are three favorable and five unfavorable outcomes in the three tosses of the coin. Using the previous reasoning I would conclude that a $3 bet on two heads and a tail, independent of the order in which they occur, should return $5, if successful. This would appear to be a fair wager, because the number of dollars involved is proportional to the number of possible outcomes.

There are two separate concepts here. One has to do with the total number of distinct outcomes independently of the order of occurence. This leads to the notion of probability. The other has to do with the additional feature of ordering and is related to the notion of correlation between outcomes. Probability and correlation are primative statistical concepts and their understanding is fundamental to identifying patterns in the measurement of phenomena.

How does this notion of a fair wager compare with reality? Consider the outcome of the 10 independent coin tosses depicted in Fig. 1.2(a). In this figure, I employ a tool for transforming a qualitative sequence, such as the heads and tails, into a quantitative sequence. Many people find it difficult to hold a string of abstract symbols in their heads, such as presented in an algebraic argument, so their concentration falters in a long tortuous logical presentation. This is sometimes referred to, somewhat callously, as attention deficient disorder (ADD), meaning that the presentation is not of interest to the person being diagnosed and so their attention wanders. Many people prefer a bottom-line presentation, with only the conclusions given. One analogy is a geometric rather than an algebraic presentation of an argument. This is done in Fig. 1.2 by replacing a head with $+1$ and a tail with -1 and adding the sequence of plus and minus ones together so that the fifth value is the sum of the first five outcomes. For example, after two tosses I replace HH with $+2$, TT with -2 and HT or TH with 0; after three steps I replace HHH with $+3$, HHT with $+1$, HTT with -1 and so on. Connecting the sequence of points defines a path or a trajectory, one path along the network in Fig. 1.1 is replaced with a single curve. In Fig. 1.2(a) the first toss is a tail followed by another two tails to yield -3. The fourth toss is a head

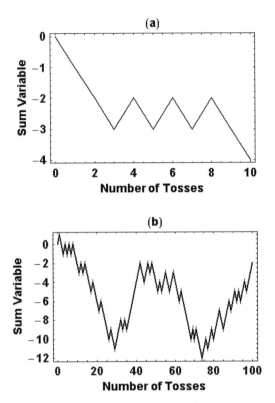

Fig. 1.2: In a sequences of coin tosses each head is replaced with +1 and each tail with a −1 and the sequence of +1 and −1 are added together to form a trajectory. (a) Here the first toss is a tail followed by a head, then a tail three times in a row and so on for 10 tosses. (b) Here there are 100 tosses and the trajectory looks quite different from the one depicted in (a) for 10 tossses, but this is due to the increase in the number of tosses.

followed by a tail, followed by a head followed by a tail and finally a head followed by two tails. In this way the sequence of tosses can be followed. In Fig. 1.2(b) the trajectory for 100 tosses is shown. At the end of this complicated sequence, the person tossing the coin has presented three more tails than heads.

All the trajectories constructed using the numerical values for the heads and tails have the interesting property that the final point uniquely detemines the difference in how many heads and tails occur in a sequence of tosses. This property of the sequence follows from the specific assignment of the numerical values of +1 to H and −1 to T. Therefore, rather than betting there will be the same number of heads and tails in a long sequence of tosses, I could make the equivalent wager that the end point of the trajectory is zero; zero resulting as a consequence of the fact that there are as many plus ones as there are minus ones. If there is one more head in the sequence than there are tails, then the final point of the trajectory will be $+1$, independently of the path taken. If there are two more heads in the sequence than there are tails, then the final point of the trajectory will be $+2$, independently of the path taken; and so on. So we have two separate pieces of information; one regarding the number of times a specific event occurs and the other is the point in time at which that event occurs.

The rather innocuous observation, that in a very long sequence of independent tosses, the number of heads and tails should be equal, is the foundation of statistics and probability theory. This convergence in the relative number of tosses onto one-half is known as *Bernoulli's Theorem,* leading eventually to the *Law of Large Numbers.* One way to express this result is through the ratio of the number of heads to the total number of tosses, as the total number of tosses becomes larger and larger. This ratio approaches one half as the total number of tosses becomes very large and leads to the notion of a probability based on the concept of relative frequency. If the relative frequency of a head occurring in a long sequence of tosses (ultimately an infinite number of tosses) is 1/2, then the probability of a head occurring in any given toss is also 1/2. This may be a rather difficult concept to wrap your head around. Let me see how we can use this relative frequency concept applied to a long data sequence. Since now I only want to count the number of heads in the data sequence, I replace T with 0 and H with $+1$. I use a different quantitative replacement of the symbols H and T, because my purpose is different from what it was before. Stay loose and don't let the details intimidate you.

In Fig. 1.3, I graph the number of heads divided by the total number of tosses (relative frequency of the number of heads in the sequence), obtained by having the computer generate one thousand

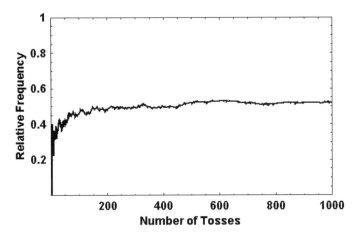

Fig. 1.3: The ratio of the number of heads in a given sequence of coin tosses is shown, up to a total of one thousand tosses. The numerical value of 0.50, that is approached with an increasing number of tosses, is the theoretical relative frequency of a head appearing in an infinitely long sequence of coin tosses. The relative frequency is interpreted as the probability for a head to occur on a single coin toss.

coin tosses. It would be time consuming to toss a real coin one thousand times and record the number of heads. Using the computer, it is a simple counting of the number of favorable events relative to the total number of events. Note that even after one thousand tosses the relative frequency of heads is still not exactly 1/2, but it is approaching that value. This is both the strength and weakness of the probability calculus and statistics: a prediction is never exactly right, but then again, it is never completely wrong. This is also what makes statistics useful in characterizing the human condition; complex phenomena are unpredictable.

Here the concept of probability is implicit in the notion of relative frequency. The fact that the probability of tossing a head is one-half, implies that the probability of tossing a tail is also one-half. One result implies the other, because there are only two possible outcomes of a coin toss and the probability of a head occurring plus the probability of a tail occurring must equal one. The sum of the two probabilities must equal one, implying that either outcome occurs with absolute certainty and that is unquestionably the case in tossing a coin.

The conventional interpretation of probability theory is that in each and every coin toss, the probability of a head occurring is one-half, the relative frequency, regardless of the number of times the coin has been tossed. For example, if I toss the coin twice, the probability of obtaining a head in the first toss is 1/2 and the probability of obtaining a head on the second toss is also 1/2, so that the total probability of getting two heads in a row is the product of the separate probabilities $(1/2)(1/2) = 1/4$. In the same way, the probability of a H followed by a T in two successive coin tosses is also given by 1/4. Thus, using the relative frequency concept, the probability of any particular sequence, or any particular path in Fig. 1.2, consisting of n two-valued (binary) choices of H (+1) and T (−1) occurs with probability $1/2^n$, with a factor of 1/2 occurring for each of the n choices made along the binary path. If there are 3 tosses, 3 binary choices, the probability is $1/2^3 = 1/8$; for 4 tosses, 4 binary choices, the probability is $1/2^4 = 1/16$; and so on. This is the standard argument, but is it reasonable?

If asked, many people would say that the probability of obtaining H after tossing the sequence HHHHHHHH is less than one-half. To such people it does not seem reasonable that after a long sequence of identical outcomes, that the probability of a subsequent outcome should still be the same; should still be 1/2. But this is what we mean when we say that one toss is independent of all the other tosses. The past does not influence the future in this view of the world. The relative frequency view of probability and the independence of sequential events is often presented as if it were obvious, but that is not the case. In the 18th century, two of the leading mathematicians of the day had a public argument regarding the efficacy of inoculation against smallpox, and the controversy centered on the meanings of probability, statistical expectation values and the utility of those ideas in the domain of social interactions and health policy.

1.4. Odds against smallpox

In the middle of the 18th century, Europe fought against a myriad of diseases with which contemporary medicine could not cope. One such disease was smallpox. The English gave the name "The small pox" to the disease because at certain stages it resembled "The great pox," syphilis.[4] It is estimated that around 1750 smallpox claimed the lives of 10% of the populations of England and France, without regard for age or station in society. Smallpox killed approximately one-third of those infected and left the remainer scarred for life, but in some outbreaks the percentage that died was far greater. The Roman empire lost 5 million in an epidemic in A.D.180, from which, Fenster[5] conjectures, the empire never recovered.

Although smallpox was one of the scourges of 18th century Europe, its prevention was known elsewhere in the world. In Turkey there was a tradition of innoculating the population against the disease, one that had been handed down from the antiquity of the Middle East, the Caucasus and Africa. The process was simple, which is not to say that it was readily acceptable to the European mind, since it came out of the folk medicine of a backward part of the world. The Turkish method was to impregnate the skin with puss from live smallpox putules from which most receipients would subsequently become immune to the disease. This was done to Turkish children once a year by the elderly Ottoman women of a city, such as in Constantanople.

Lady Mary Montagu, a well known member of the English aristocracy, was instrumental in engineering the political acceptance of this inoculation technique into Europe. Once a beauty and a much sought after member of society, she contracted the disease, survived and was left scarred for life. In this way she acquired first-hand knowledge of the disease and knew its deadly nature. Her husband was ambassador to Turkey and while there she observed the Turkish method of innoculation for herself. She became so convinced of its efficacy that she had her five-year-old son inoculated while the family was in Constantinople. In 1718, she brought the detailed knowledge of the procedure back to England and eventually convinced Caroline, the Princess of Wales and future Queen of England, to have her two children inoculated, which she did in 1721.[5]

What makes this deadly disease interesting in the present discussion is that a small percentage of those being inoculated contracted the disease and died; consequently becoming inoculated was a gamble. The odds in this life-and-death wager were one in seven that a person would contract smallpox at some time in their life without innoculation, versus a one in two hundred chance of dying from smallpox within a few months of being inoculated.[6] Thus, one had to balance the long term good of an action against the short term risk of that action. The public controversy over inoculation concerned the applicability of everyday experience to the idealized reasoning used in gaming and probability theory.

Daston[6] relates that the European intellectuals, which at that time meant the French, and in particular Voltaire, embraced the idea of inoculation against a disease. However, mathematicians sought to separate the arguments from their emotional trappings and determine if the benefits of inoculation against smallpox could be determined by the application of reason alone. The two antagonists in our little drama were Daniel Bernoulli (pro) and Jean de la Rond d'Alembert (con), who argued for and against the rational basis for inoculation, not about inoculation itself. They were both mathematicians and in the real world each supported inoculation against smallpox.

Daniel Bernoulli was second generation in a family scientific dynasty that lasted 150 years (see Fig. 1.4). Daniel's uncle, Jacques (Jacob), was the first of the Bernoulli's to gain a reputation in mathematics, being born in Basel, Switzerland in 1654 and becoming a professor of mathematics in 1687 at the University of Basel. Jacques wrote the second book ever on the theory of probability that was published posthumously in 1713. Jean (I) was the father of Daniel, the

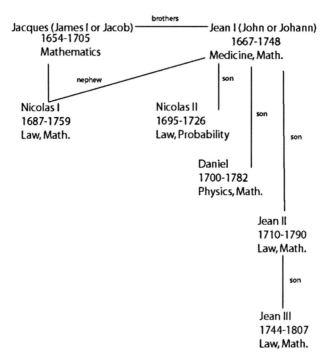

Fig. 1.4: This is the 150-year family tree of the Bernoulli family; arguably one of the most influential scientific families in history.

brother of Jacques and although 13 years younger than his brother was his lifelong scientific rival. Both Jacques and Jean (I) studied with Leibniz, the co-inventor of the calculus with Newton, and through their correspondence with Leibniz obtained many important mathematical results. In fact, much of what is contained in an elementary course on the calculus can be found in the correspondence of the two brothers and their mentor. In 1705, with the death of his brother, Jean (I) was elected to fill his brother's position in the University of Basel. It is often difficult to separate the mathematical contributions of the two brothers.[7]

Daniel was born in Groningen, Holland (1700–1782) while his father Jean (I) was professor of mathematics at the University of Groningen. Daniel studied under his father with his brother Nicolas and fellow pupil Euler, probably the most prolific scientist who ever lived. Daniel was professor of mathematics at the Academy of Petrograd (St Petersburg), Russia from 1725 to 1733, after which time he returned to Basel, where, like his father and uncle before him, became a professor at the university. Daniel's brother Nicolas, who was with him in St Petersburg, died quite young in a drowning accident, but not before making some fundamental contributions to probability theory.[7] Daniel Bernoulli was the protagonist in the discussion of probability theory and smallpox.

D'Alembert fell into the category of the 20% of children abandoned to Foundly Hospitals in the 18th century. Jean le Rond d'Alembert, who was destined to become the greatest French mathematician of his day, was born in Paris (1717–1783) and left on the steps of the Church of St Jean le Rond, near Notre Dame.[8] His mother was a member of the aristocracy and he was the bastard son of Chevalier Destouches, who, on finding out that the child's mother had abandoned him on the steps of a church, put him in a foster home with an annual income of 1200 livers. After achieving national and international fame, d'Alembert was to spurn his natural mother and acknowledge his foster parents as his own. How or why he adopted the name d'Alembert remains a mystery.

D'Alembert first studied law, then medicine, all the while learning mathematics on his own. At the age of 23, he became a member of *Académie des Sciences* as its astronomical correspondent. His rise was rapid and by the age of 37 he was recognized as France's leading mathematician and natural philosoper. One of the major intellectual enterprises of the day was the *Encyclopédie*, which was published in 28 folio volumes between 1751 and 1772. He was joint editor with Denis Diderot (1713–1784) for the first seven years of the project, and after he had stepped down as co-editor he continued to contribute most of the scientific articles thereafter.[7] He was invited by Fredrerick the Great to join the Berlin Academy in 1746, and later by Catherine the Great of Russia to join her Court. He was the antagonist in the discussion on probability theory and smallpox.

Daniel Bernoulli published a memoir on smallpox inoculation in 1766 in the *Histoire et Mémoires de l'Académie des Sciences*, examining both its advantages and disadvantages. D'Alembert had access to a copy of the paper in 1760, what we today call a preprint, or prepublication copy, of the paper, probably because of his position as *secrétaire perpetuel* of the Académie. Whatever the history, d'Alembert gave a public presentation on the inoculation problem at the Académie on November 12, 1760, "...which was a long and detailed critique of Bernoulli's memoir on the subject."[6] The core of d'Alembert's argument was that "...mathematicians still lacked both sufficient information on smallpox inoculation and mortality rates, and also a reliable method with which to apply the theory of probabilities to such problems."[6] At the same time d'Alembert went out of his way to publicly endorse inoculation, while professionally continuing to attack the use of probability theory to support the implementation of smallpox inoculation.

D'Alembert's criticism rested on the fact that the experience of the individual was neglected in the determination of probabilities. By its very nature Bernoulli's law of large numbers excludes the variability of the individual and focuses on what is common in very large populations. This perspective, of course, carried over to his treatment of life expectancy with and without inoculation. In his discussion Bernoulli assumed that the likelihood of contracting smallpox and dying from it was the same for children and adults.

Bernoulli's failure to account for the experience of the individual was unacceptable to d'Alembert. The details and variability, that enrich the socio-medical problem, were smoothed over to emphasize the significance of the mathematical question, that being, how probability, derived from a group concept (relative frequency) applies to the individual. Looking at Fig. 1.2 it is evident that the unpredictability of a coin toss gives rise to a curve that has many changes in direction, but the trajectory also has many intervals where the curve travels up or down in a straight line indicating a "run" of either heads or tails. The notion of probability as provided by Bernoulli wipes out this variability and replaces it with a single number, as indicated in Fig. 1.3.

The controversy between the two mathematicians was more intense because the difference between the probability of an event occurring and the expectation of an event occurring was not clear. If we return to gambling we can see that there is a difference between the probability of one's winning and how much one expects to win. For example, suppose we bet on three heads occurring on three successive tosses of a coin. We know that the probability of three heads occuring is $(1/2)(1/2)(1/2) = 1/8$, but if we bet ten dollars, how much do we expect to win in a long sequence of such bets?

The notions of probability and expected value are closely related. In gambling parlance the expected winings would be the expected return on one's wager. The expected return is defined mathematically as the size of the bet times the probability of winning. For example, if I wager two dollars on head at the toss of a coin, and since the probability of heads occurring is 1/2, my expected return is one dollar. This does not mean that I will receive one dollar for each and every bet; in a single bet I will either win two dollars or lose two dollars. Therefore the expected value does not have any connection with a single wager and only takes on meaning for a large number of wagers. Thus, for a long sequence of bets a fair wager would be one in which I put up one dollar in order to receive back two dollars if a head comes up on the toss. But this convoluted logic may not be clear.

Consider a more elaborate game. Suppose, there are 100 top hats on a table. Out of the 100, a given amount of money is placed under only 20 of the hats, leaving 80 with nothing. Under 15 of the 20 hats is a $10 gold piece each; under four hats is a $50 gold piece each; and under the last hat is a $100 gold piece. I am now asked to choose one of the 100 hats. What is my mathematically expected gain? How much do I expect to win? As mentioned above, the expected gain is given by the product of the amount of money under the hat and the probability of a given hat covering that amount of money. Therefore, using the relative frequency notion, the probability of choosing a hat covering no money is 80/100, that for a hat covering $10 is 15/100, that for a hat covering $50 is 4/100 and that for a hat covering $100 is 1/100. Using these values of probability and gain, my mathematical expectation is

$$\$0 \times \frac{80}{100} + \$10 \times \frac{15}{100} + \$50 \times \frac{4}{100} + \$100 \times \frac{1}{100} = \$4.50.$$

Consequently, in a sufficiently long sequence of hat selections, I would expect to gain $4.50 per game, on average. As Weaver[9] points out in his treatment of a similar example, a large fraction of the time I would not receive anything (about 80% of the plays). In about 15% of the plays I would receive $10; in about 4% of the choices I would receive $50 and in about 1% of the selections I would receive $100. In order for the theory of probability to work I must play for a long time and make many selections.

Fairness comes into the process by requiring an ante. In order to play the game it is reasonable for me to ante $4.50 each time before I choose a new hat. The ante is equal to my expected winnings and that is what makes the wager fair. If I play the game for a sufficiently long time I will win slightly more money than I lose and on average I will win $4.50 per play. A gambling house would charge a little more than my expected winnings of $4.50 to allow me to play the game; the amount over $4.50 would be their margin of profit over the long term. I might be willing to pay the overcharge for the thrill of playing and knowing that without a slight edge the casino cannot stay in business.

Of course there is an alternate view, in which I can look at the result of the selections of a large number of people, each making a single choice. There is a perceived equivalence between one person making a large number of choices and a large number of people each making a single choice. In one case the expected winnings is determined by the average over the sequence of individual choices and in the other by the average over the ensemble of single choices. The expected winnings should be the same.

But how do the ideas of probability and expectation value figure into the argument over smallpox inoculations? D'Alembert argued that one would need an acurate mortality table to determine how long a person is expected to live if inoculated or if s/he remained inoculation free. Without such tables, D'Alembert reasoned, Bernoulli could not determine if the life expectancy increased or decreased by being inoculated. D'Alembert gave the example of two theoretical mortality curves, similar to those depicted in Fig. 1.5, from which we see that the number of persons suviving to a given age changes with age. The solid curve displays a population that stays nearly constant for a time and then begins to die off at a significant rate, until they are all dead. The population of the young is substantially larger than that of the old. On the other hand, the dashed curve displays a population that dies off quickly at young ages, levels off at some intermediate age, and then again dies off somewhat more quickly at old age. The second mortality curve favors the older, whereas the first mortality

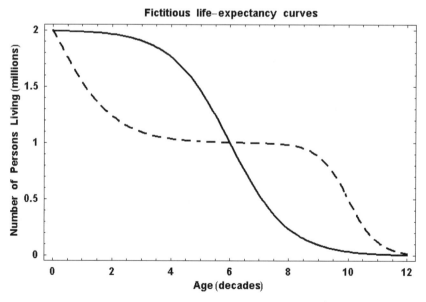

Fig. 1.5: Life-expectancy curves in the manner of d'Alembert, where the unit values on each axis are taken for convenience and are not intended to portray any actual population.

curve favors the younger. But which mortality curve corresponds to reality?

The two curves are quite different from each other and the life experiences of the two populations are quite different, but the average age in the population of the two curves could well be the same. I could spend a lot of time interpreting the two curves. For example, for the dashed curve babies and children die at a higher rate than do the adults, perhaps suggesting a Darwinian interpretation of the society. The young and weak die off relatively quick and those that survive live for a long time. For the solid curve mature adults and seniors die at a higher rate than do the children, perhaps suggesting war and a failure for society to honor its elderly. But as much fun as such speculation is, d'Alembert's point was that since the mortality curves were not known, the societal influence of inoculations on life expectancy could not be reliably predicted.

However, d'Alembert was not accurate with regard to his position on the existence of mortality tables, even if he was correct about Bernoulli's failure to use such tables in the determination of the smallpox inoculation argument. The Romans were concerned with life expectancy, because annuities were commonplace and the equivalent aggregate value of such annuities had to be computed, so they needed to know how long people were expected to live. As pointed out by Montroll and Badger,[2] life expectancy tables for this purpose were proposed by Praetonian Praefect Ulpian and Jurisconsult Aemilius Macer around the 3rd century A.D. The life expectancy as a function of age given in these tables is summarized as one of the curves in Fig. 1.6. A similar table was constructed by Edmond Halley based on the mortality experiences in Breslau during the period 1687–1691 and is the second curve in this figure.

Figure 1.6 shows clearly the influence of the progress in medicine and public health on life expectancy over a 1500 year period. The bump in the Halley's curve for ages less than 20 is a clear indication of the survival of children. Consequently, as bad as the conditions for children in 17th century Europe seemed to be, they were vastly

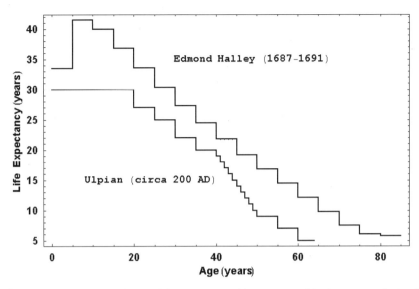

Fig. 1.6: The data from the two life expectancy tables mentioned in the text are depicted. The lower one is of Roman origin and is nearly 2000 years old. The upper one is nearly contemporary being only 300 years old.

superior to the situation faced by the children of Rome. Of course such tables are updated frequently in modern times, with the peak in the distribution of population shifting towards the age at which individuals collect social security. These curves, or rather ones just like them for the modern world, are used by governments to determine their budgets for those being supported on various social security systems. The mortality curves are also used by insurance companies to determine survivor benefits. Both these calculations would be strongly influenced by medical breakthroughs such as finding a cure for cancer. The insurance companies would not have to pay survivor benefits, consequently their profits would increase. On the other hand, the government would have to support a retiree for a longer period of time, thereby burdening the remaining workforce. Thus, Bernoulli could have taken the variation of mortality with age into account in his arguments. But putting this aside, d'Alembert's criticism of the sociological application of probability went even deeper.

D'Alembert questioned the fundamental notion of mathematical expectation, since to him, such expectation ran counter to the actions of reasonable people. For example, suppose there are two wagers that one could make, putting at risk the same amount of money. In one case there is a 99% chance that a bet would return $1,000 and in the second case there is a 0.1% chance that a bet would return $990,000. He argued that a reasonable person would always choose the more certain bet, and therefore the expected return is not the same in the two cases, even though their mathematical expectation of $990 is the same.[6] He argued that it is prudence and the psychology of risk that are not properly taken into account in the determination of probability as Bernoulli applied it to human behavior.

For the mathematical establishment, as represented by Bernoulli, the calculus of probability represented the perfect reasoning of a reasonable person in the absence of complete knowledge. For them, mathematics was a prescriptive way of enabling the rest of humanity to reach proper conclusions that they could not reach if left to their own devices. For d'Alembert, the central flaw in the calculus of probability was that it ought not to be a measure of ignorance concerning our understanding of a situation, but should instead be a way to describe the psychology of risk for the common person. One camp believed that the calculus of probability prescibed the ideal to which humanity should aspire and the other believed that it ought to describe the reality of ordinary people. History has endorsed the mathematical ideal over the psychological reality, and many of today's social problems can be traced back to this choice. In particular, I argue that the use of the expected value and the suppression of variability in the description of medical phenomena has blinded generations of scientists to sources of information that was readily available to them. This information was discounted because it lacked regularity and varied too rapidly, resembling more the warming up of an orchestra than their playing of music.

1.5. Information and chance

Gambling was compatible with the culture of the 1700s in part because the agrarian population was close to nature and therefore people knew that in spite of their best efforts the world could be beyond their control. The caprice of weather could destroy a farmer's crop overnight; a disease such as smallpox could lay low an entire family in a matter of days and misfortune could put one in debtor's prison. Gambling was in conflict with the teachings of the church, but not with the experience of a population that could be decimated by a variety of plagues. The devastating diseases of the 17th and 18th centuries, to those that survived them, apparently chose its victims in ways that eluded reason. When reason no longer explains events the mind embraces either supersition or faith.

The stability and order sought by the general population, since they could not rely on nature, came through the church and the law. The two were not always in harmony; but the sporadic conflict between theological and secular dogma only sharpened the need for society to have explicit laws that clearly established the doctrine of fairness. Every gambler knows that if the amount to be won is proportional to the wager made, and everyone wagering has the same proportion, then the bet is fair. People could understand what it meant to be given a fair chance at life, at least in the abstract. To attain this fairness in society individuals were willing to give up most of what made them different from one another. This is not to be confused with equal opportunity under the law, which is a concession that society makes to the individual, whereas equality under the law is a concession the individual makes to society and it is only fair.

In the 17th century, as now, the dominant characteristics of being human were curiosity and anxiety, specifically about the future. As people became more urbanized, the window into the future became greater and the attempt to anticipate what will happen for longer and longer periods of time became more ingrained into the society. In the

18th century people apprenticed their children; today parents plan for their children's education, from preschool through university. Now, as then, the purpose of the planning is to make the future more certain, to suppress the unexpected, and to insure that our children live long and harmonious lives. Perhaps even in the caves, mothers fought to get the largest portion of the day's kill for their offspring.

In this chapter I have introduced a number of new concepts based on probabilitic reasoning. But unless you are mathematically inclined all this talk about coin tossing may leave you with an uncomfortable feeling that I am trying to put something over on you. It is therefore probably worthwhile to present the discussion in a different way, a way that replaces probability with information. Information serves the secondary purpose of providing a quantitative measure of the complexity associated with a given phenomenon. Claude Shannon co-invented information theory in 1951 with Norbet Wiener, or at least that was when his first paper on the subject was published.[10] They used an analogy with thermodynamic entropy to quantify this newly coined term. It was well known that entropy provides a measure of order in physical systems, so they reasoned that the same expression, in a more general context, could be used to measure order in social systems.

The information measure quantifies the level of surprise associated with the outcome of an experiment given a certain number of alternative outcomes. In the case of coin tossing the same situation arises as in the binary code used to store information in a digital computer. It is probably familiar to the reader that deciding between 0 and 1 in a binary code corresponds to gaining one bit of information. Consequently, when I flip a coin, I also gain one bit of information, when I see the head or tail appear. A string of n binary alternatives, that is, either zeros or ones, or equivalently Hs or Ts, yields n bits of information. Note that in terms of probabilities, a string of n binary alternatives is given by the product of 1/2 by 1/2 a total of n times, so probabilities are multiplicative. On the other hand, information is additive, each outcome in the string yields one bit of information.

Consider the binary tree given by Fig. 1.1, each path along the tree is a particular combination of three outcome of Hs and Ts, so I gain three bits of information along each path into the future.

Information Theory therefore provides a transformation from the multiplicative behavior of probabilities to the additive behavior of information. In a binary system I can determine how much information is gained by making the probability of the two outcomes unequal. Let me use a weighted coin so that the head comes up more frequently than does the tail. The information associated with this unfair process is drawn in Fig. 1.7 as a function of the probability

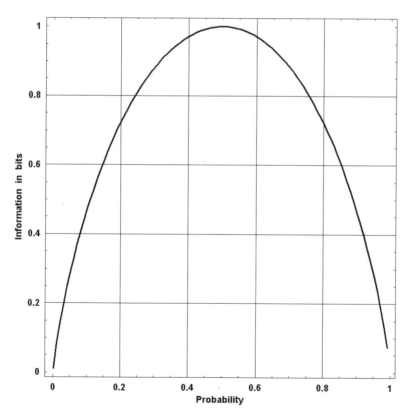

Fig. 1.7: The vertcal axis is the information gained in bits by tossing a biased coin. The horizontal axis is the degree of bias measured by the probability of one outcome in a binary process.

of a head coming up on a toss. In fact the figure is drawn for all possible weightings of the coin, making it either more or less likely to come up heads. It is evident from the symmetry of the information curve that the maximum information gain per toss is associated with a fair game, that is, one bit of information is obtained when the outcomes are equally likely and the probability of each outcome is one-half. Any other configuration results in less information being gained with any given outcome. A little reflection will show this is reasonable, since if the coin is weighted in favor of a head we are less surprised when the head appears. This measure also emphasizes the importance of a fair process, one in which it is preferred to have the maximum information gained in each realization.

Note that it is possible to use information to characterize a random walk process, since with each step the walker moves either to the right or to the left, so there is one bit of information per step. For a sequence of steps I can construct the total information by adding one bit per step and the total number of steps gives the maximum information associated with the path of the random walk. On the other hand, if the walk were biased then I could use Fig. 1.7 to determine how much information could be obtained from a single realization of the random walk. Suppose the probabilty of stepping to the right is 3/4 and stepping to the left is 1/4, then from Fig. 1.7 it could be determined that the information gain is 0.81 bit per step instead of one bit per step. Consequently in a walk of N steps the information gained, rather than being N bits, is $0.81N$ bits, the fact the information is less in the biased case should come as no surprise.

Chapter 2

The Expectation of Health

A concerned mother takes her child's temperature to determine if he has a fever. A physical therapist monitors her client on a treadmill looking for signs of weakness and/or recovery. The nurse carefully releases the air from the cuff to determine the systolic and diastolic pressure of her charge. The physician counts the number of beats in 10 seconds and multiplies by six to obtain his patient's heart rate in beats per minute. These are the coarse, preliminary measures that medical practioners use to characterize health: body temperature, gait interval, blood pressure and heart rate. There are certainly a number of other measures as well, but I shall restrict most of my discussion to this handful. What I have to say about these quantities are, in fact, equally true of the other measures as well.

It might be said that each of these individuals, the physical therapist, nurse, and physician, has the same medical view of how the body operates. A view that is remarkably consistent with our technological society in which the temperature of our homes is controlled by a thermostat, our television and lights can be turned on or off with the sound of our voice, and the speed of a car on the highway is determined by cruise control. It is not clear when this view of how the body works began to permeate society, but in medicine the concept was introduced by the 19th century French scientist Claude Bernard (1813–1878). He developed the concept of homeostasis in his study of stability of the human body. In his book *Introduction to Experimental Medicine* (1865) Dr. Bernard conjectured:

> Constancy of the internal milieu is the essential condition to a free life.

The word for this behavior, homeostasis, was popularized half a century later by American physiologist Walter Cannon (1871–1945) in his 1932 book *The Wisdom of the Body*[11] and is the conjoining of the Greek words for "same" and "steady." This notion of homeostasis is what many consider to be the guiding principle of medicine. This is the view I challenge with the arguments in this book and using the work of my collegues from the corresponding literature.

Every human body has multiple automatic inhibition mechanisms that suppress disquieting influences, some of which can be controlled and others cannot. Homeostasis is the strategy, the culmination of countless years of evolution, by which the body reacts to every change in the environment with an equilibrating response. The goal of each response is to maintain an internal balance, which evolution has decided is a good thing for animals, although it is not always evident how a particular response is related to a specific antagonism. However, in many cases this connection can be made explicit, for example, the size of the pupil of the human eye changes inversely with the intensity of light entering the retina; the greater the intensity, the smaller the pupil and vice versa. This physiological balancing occurs because too much light will destroy the light sensitive cones of the retina[12] and the person will be blinded. This is how we interact with our environment.

Homeostasis can therefore be seen as a resistance to change and may be viewed as a general property of all highly complex systems that are open to the environment. Homeostatic systems exist, for example, in ecology, where a species can be robust or fragile, depending on the balance in the complex food webs. A fragile species might become extinct if the external changes are too great, or if the external disturbance is unique and is therefore one for which no evolutionary response had been selected. The stability of the system as a whole is required to maintain the ecological network, but the extinction of one or more species within the network may be the price of that overall stability. This conclusion assumes that the disturbance from the environment persists over time, say, in the over-fishing of a particular region of the ocean. However, if the perturbation is suppressed, say by restricting the time of the year these regions may be fished, the system may relax back to its original dynamical balance. On the other hand, there are conditions of delicate balance within some systems where even the smallest changes can have dramatic consequences. I examine such systems in due course.

Biology not only views humans and other animals as homeostatic systems, but microscopic cells, as well. The chemical reactions within a cell and its interactions with its environment are certainly as complicated as an ecosystem, if not more complicated. In sociology, from a certain perspective, industrial firms and other large organizations such as universities may be viewed as homeostatic systems, that are resistive to change. In all these examples the system maintains its structure and function through a multitude of interdependent nonlinear dynamical processes acting as regulators. It is this view of regulation and control that I need to examine and understand in order to introduce an explanation of the stability of complex physiological systems that has emerged over the last century or so.

Biology teaches that evolution has, over the years, reduced homeostatic systems to the bare minimum, so that in the spirit of parsomony, every internal mechanism of a biological system is necessary to maintain either the structural or functional integrity of the organization. Homeostatic physiologic systems are therefore ultrastable, regardless of the number of switches. However, such stability is only true for a given range of parameter values in complex systems. The intrinsic nonlinearity of homeostatic systems makes their long-time behavior unpredictable, as every CEO of a large company knows.

It is profitable to examine the difference between the processes that enter into the thermal regulation of physical systems and those that control physiological systems. This discussion is useful because it focuses thinking on what is different in these two ways of regulating physiologic indicators such as temperature. In particular such regulation demonstrates that homeostatic control is completely alien to Newtonian dynamics and the homeostasis idea is used to examine how well the fluctuating behavior of physiological time series can be explained. Here I use breathing rate, tidal volume and heart rate as three exemplar time series. Simple arguments show that the fluctuations in the rates cannot be explained as uncorrelated random noise, as would be expected if the underlying phenomena were homeostatically controlled in the traditional sense. Instead, faint patterns in the fluctuations are found that remain to be explained. However, that explanation must wait a few chapters in order to develop some preliminary concepts.

2.1. Control and cybernetics

Why should physiologic systems be homeostatic? Why has nature determined that this is the "best" way to control the various complex systems in the human body? In part, nature's choices have to do with the fact that no physiologic system is isolated, but is, in fact, made up of a mind-numbing number of subsystems, the cells. The task of a cell is simple and repetitive, but that of an organ is not. Therefore a complex system like the lungs is made up of a variety of cell types, each type performing a given different function. If response to change in the external environment were at the cellular level, physiology would be much more complicated than it already is, and organs would no doubt be unstable. But nature has found that if the immediate environment of the cells is held constant, or kept within certain narrowly defined limits, then the cells can continue to perform the tasks for which they were designed and no others, even while an organ responds to the external disturbances. As long as the internal environment stays within a certain operational range, the cells will continue to function without change. Thus, homeostasis is the persumed method that nature has devised to keep the internal state of the body under control.

The body can only operate within a fairly narrow range of parameter values, for body temperature, pH, oxygen concentration and carbon dioxide levels in the blood, blood pressure, muscle tension, hormone concentration and on and on. But maintaining these quantities at specific values requires sophisticated control mechanisms. Therefore before considering these systems I introduce the nomenclature of homeostasis and control through the example of driving a car at a fixed speed along a highway.

Driving a car along the highway is a complicated business, so to be specific, suppose I am on the California Coast Highway #1, between San Francisco and Los Angeles, driving south. The speed at which I drive the car is determined by a number of factors, including hairpin curves in and out of the mountains, rocks across the road, traveling up and down hills, and the frequency at which small towns appear along the road. I increase and decrease my speed as necessary, but I always come back to what I had decided is the optimal speed for the trip, one usually related to the speed limit. This optimal or predetermined speed is called the *set point*. It is the speed of the car that is under my control, so in the unimaginative way scientists name processes, the car's speed is called the *controlled variable* or the control variable.

In an abstract way I adjust the control variable to counteract changing environmental conditions, but always with a view to returning to the set point. But I need to know the instantaneous speed of the car in order to adjust the speed and for this I need a measuring device. In a car the measurement of the control variable is displayed on the speedometer. The measurement of the car's speed is done indirectly by a sensor that counts the number of revolutions of the tire in a fixed time interval, then knowing the size of the tire, the distance covered in that time interval can be determined as well as the speed.

The data gathered by the sensor is transmitted to the speedometer where it is transformed into information and displayed. I can see the display and use the information. The connections carrying the data, from the sensor where it is detected, to the site where the information can be used, is called the *afferent pathway* or simply the afferent (see Fig. 2.1).

The data from the sensor is not useful unless it can be transformed into information and similarly the information is not useful until it is transformed into knowledge. I need to know how the present speed compares to the set point in order to decide what to do; increase my speed or decrease my speed. The decision-making process that determines what I do to control the speed of the car occurs in my brain, so that the brain is called the *controller*.

This is the first half of the control process. The sensing of the speed, the transmission of data to the display, the interpretation of the information and making the decision of what to do. But once the decision is made, that information must be passed back down to the point of implementation. The implementation of the decision is done

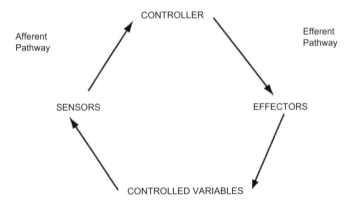

Fig. 2.1: The basic ingredients of a homeostatic system are indicated. The information is passed from the controlled variable along the afferent pathway by means of the sensors to the controller. Once the controller has made a decision on what to do, information is passed along the efferent pathway from the controller to the effectors, thereby changing the controlled variables. This cycle, functioning in a negative feedback mode, maintains the control variable at the set point.

through effecting the value of the control variable by pressing on the brake or accelerator, the so-called *effectors*. The nerve impulses from my brain, travel down the central nervous system, to the peripheral nervous system to my foot. The pulse tells my foot to step on the brake or accelerator and this is how the decision is communicated from the controller to the effector. This trail is the *efferent pathway*, or efferent.

Two configurations have to do with the way in which information is used in a control system. For example, the present speed of the car is determined by whether I have just accelerated or braked and I use this information along with the present speed to determine what to do next. Thus, I reflect back on what has previously occurred and use this knowledge to determine future decisions; this is called *feedback*. Physiologic homeostatic systems are often referred to as *negative feedback* systems, because systems respond in ways to dampen any environmental disturbances, but this is not the complete story. Certain systems have the opposite behavior, that is, they have a *positive feedback*, because the systems respond in ways to amplify environmental perturbations. Of course, such a response would lead to unstable behavior in general, but such instabilility is sometimes useful. Consider, for example, the situation of labor during childbirth. The contractions of the uterus eventually force the baby's head into the cervex. The stretching of the cervix, to accommodate the baby's head, induces the release of oxytocin. The increased level of oxytocin induces increased contractions of the uterus. The increased contractions force the baby's head further into the cervix. The contractions become stronger and stronger, through the process of amplification, until the baby is born. Childbirth is a clear example of positive feedback. Of course the process must shut off once the baby is born.

Another example of positive feedback is the spreading of an epidemic. If one person with a virus enters a room full of people, where the virus is not present, then as the virus is spread through the air (or alternatively by contact), other people become infected. Each of the infected individuals becomes a new source of infection. Eventually the entire room of people will be infected with the virus, for a party lasting a sufficiently long time; positive feedback produces explosive growth which will only stop when everyone is either infected or immune. This is the mechanism that produced the devasting effects of smallpox and other diseases throughout human history.

Consequently feedback can either amplify or suppress disturbances depending on the system's past behavior. In a complementary way I can anticipate what might happen in the future, which in the car would be through the use of road signs, the behavior of traffic and the weather. I use information about what lies ahead to minimize the influence of those conditions on the control variable when I eventually encounter them. This anticipatory use of information is called *feedforward*. Feedforward and feedback are essentially the two main modes of operations of a homeostatic system. In complex hierarchial systems, low-level feedforward and high-level negative feedback act to constrain system growth generated by low-level positive feedback.

Control is the central concept of Norbert Wiener's (1894–1964) discipline of *cybernetics* introduced into the scientist's lexicon in 1948 through a book of the same name. Here again the word is the concatenation of two Greek words, the prefix for "to pilot" and the suffix for "to steer," so that it literally means "steersman." As a discipline cybernetics[13] attempts to determine how and why natural control systems seem to be more versatile and adaptabile than are their mechanical man-made analogues. As first envisioned by Wiener, cybernetics was the science of control and communication in both animal and machine, but it has come a long way since then. Without going into detail, cybernetics has to do with the control of complex phenomena through the construction of sophisticated models that are able to maintain, adapt and self-organize to carry out their function. In fact, cybernetics initiated the examination of the same problems that today are studied under the rubric of *complex adaptive systems* (CAS), the difference being the goal orientation of cybernetics, that is not always present in CAS.

Cybernetics[13] is interesting to consider because it allows scientists to formulate ideas that would be anathema to them using Newtonian physics. If I follow the logic of control, the effect is fed back to the controller, or cause, that modifies the effect, that is then fed back into the cause in a cyclic fashion. Therefore the simple cause-effect chain that is characteristic of Newtonian dynamics, is lost in the circular logic of feedback. In order to demystify this idea, consider the beating of the human heart; this, too, is a circular feedback system that defies linear Newtonian causation. This kind of circular logic has always been excluded from the physical sciences because it leads to the logical paradox of self-reference. An example of such self-reference is the statement: "This statement is false." If I start from the condition that this statement is true, then I immediately see that the statement is false. Given the condition that the statement is false, I see immediately that the statement is true. The subjective truth of the statement oscillates between true and false, never coming to rest. Rather than be repelled by such circularity, cybernetics embraced the feedback loop as a way of understanding such fundamental phenomena as self-organization, goal-directedness and identity. It was recognized that such circular processes are ubiquitous in complex natural phenomena.

This idea of self-reference and the contradictions it engenders lies at the heart of a world-shattering mathematical proof given by Gödel in the middle 1930s. Of course, it was the world of the mathematician that was being shattered. At that time the internationally known mathematician Hilbert was the champion of the idea that everything in mathematics could be derived from a set of axioms, like those we learned in the high school geometry of Euclid. Hilbert had the reasonable expectation that given a number of assumption (axioms) one could prove the truth of all the statements that can be made regarding a mathematical system, whether the system is geometry or number theory or anything else. Then along came Gödel to upset everything. Gödel *proved* that one could not prove the truth of all possible statements made within a mathematical system using only the axioms from within the system. Thus, in one theorem the complete program of mathematics that Hilbert had envisioned was destroyed. The incompleteness theorem of Gödel implies that scientists can never be absolutely certain about the truth of all the statements in a mathematical system and therefore by extension they can never be absolutely certain about the fundamental mathematical structure of any scientific model of reality, such as physics. Consequently, a sentence, such as "I am an invalid statement" is true only if false and false only if true. This is a perfectly formed grammatical sentence, but determining its "truth" might give a reasonable person a headache; but then mathematicians can be very unreasonable.

The human brain can retain the paradox of self-reference, simultaneously holding the same sentence to be both true and false, but the algorithmic logic of the digital computer cannot. Thus, the fact that physiological phenomena may contain such paradoxical behavior suggests that the simple models used to describe them are limited. Therefore I will review certain regulatory processes in order to distinguish between physical and biological phenomena and the models used to understand them.

2.2. Temperature regulation

The temperature of an organism is a measure of its heat content; every living organism absorbs, generates and releases heat, in order to reach an energy balance characterized by its temperature. Of course, inanimate objects have temperatures as well. A pond absorbs radiant energy from the sun during the day, stores the energy in its molecular motion, only to return that heat to the air once the sun goes down. To understand this process of cooling, visualize the air above the water as mostly empty space, but still containing gas. This gas consists of molecules of oxygen, water vapor, nitrogen and carbon dioxide whizzing around in mostly empty space and the molecules only occasionally collide with one another. The pond, on the other hand, consists mostly of water molecules that are also moving around at a speed determined by their kinetic energy, but here the collisions are innumerably more frequent because of the longer-range interaction among the particles and the greater density of the fluid over that of the air. At the surface of the pond those water molecules with a sufficiently high velocity break through the surface and escape into the air. At the same time molecules of water vapor from the air break through the surface from the other side and enter the pond. This trading of particles goes on continuously and when the net energy of the particles entering the pond is the same as the net energy of the particles leaving, over a sufficiently long time, the air-pond-sun system is said to be in equilibrium.

After the sun goes down, the air quickly cools by radiating its heat away. This is a second mechanism for energy loss. The pond takes longer to cool because the energy radiated by one water molecule is quickly absorbed by nearby water molecules. Consequently any small volume of fluid retains its average heat content, until the radiation reaches the surface and escapes into the air. Thus, the pond remains warmer than the air at twilight even though more particles leave the surface of the pond than enter it. The pond is eventually cooled at night by the escaping particles that carry energy away in addition to radiation.

An organism sometimes operates in the same way as the air-pond-sun system, but most of the time it operates differently. A sweating jogger is grateful for the phenomenon of evaporation; because it is the overheated water from her body that is escaping into the air and carrying away some of the uncomfortable excess heat. But this is only one of a number of cooling mechanisms available to the human body. These mechanisms are important because humans can only function properly when the body's core temperature is within a very narrow band.

The pond continues to absorb radiation from the sun and its temperature rises, until the rate of heat absorption matches the rate of heat loss, at which point the temperature levels off, because the energy gained and the energy lost in an interval of time are in balance. The rate of heat loss is determined by evaporation, which we have discussed, but in addition there is radiation, conduction and convection. Radiative heat loss is the energy radiated away as electromagnetic radiation, which typically covers a broad spectrum with a maximum determined by the temperature of the radiating body. The wavelength of the emitted radiation depends on the surface temperature of the radiating body, such that the wavelength gets shorter with increasing temperature. The surface of the sun is 6600 degrees Kelvin (°K) and is quite visible, whereas body temperature is approximately 300°K and its radiation is not visible to the naked eye. Night vision headgear, sensitive to infrared radiation, makes body radiation visible and enables soldiers to see one another, as well as the heat impressions of the enemy. A campfire passes through these two modes, visible when flame is present (hotter) and invisible after the flames have been extinguished (cooler), but its warmth, through long wavelength radiation, is still felt by an outstretched hand. The amount of radiant heat energy is dependent on the difference in the surface temperature of the object and the ambient temperature; the greater this temperature difference, the faster the loss of heat due to radiation. Thus, the warmth of the campfire lingers long after the flame is gone, since the ashes are only slightly warmer than the night breeze.

Finally there is the effect of convective heat loss, or heat loss due to the circulation of mass. In evaporation "hot" molecules are exchanged for "cooler" molecules at the surface of the pond. Similarly, hot molecules within the pond are cooled by radiating energy, energy that is absorbed by molecules above the pond surface. Both mechanisms heat the air above the pond and thereby slow down the cooling process by reducing the temperature difference between the surface of the pond and the air above. A slight breeze sweeps the warm surface air away and replaces it with cooler ambient air, thereby increasing the temperature difference and increasing the heat loss from the pond. In the absence of any external forcing (breeze) the warm air near the surface rises because of its reduced density and is replaced with cooler denser air. These physical processes work equally well for cooling the human body.

The human body has more than just these simple physical mechanisms available to regulate its temperature, however. The operation of one physiological mechanism is observed by the reddening of the skin when the environment becomes uncomfortably hot. The network of small blood vessels just beneath the skin controls the amount of blood flowing near the body surface. With an increase in body heat these vessels dilate, thereby increasing the flow of blood near the skin and increasing the transport of heat from the body to the environment. The increase in blood flow just under the skin produces the characteristic red color. With a decrease in the ambient temperature these same blood vessels constrict so that the blood is carried by deeper blood vessels, thereby preserving the heat in the body.[14]

Thermoregulation of the body is an archetype of homeostasis. The human body has an ideal surface temperature of $98.5 \pm 1.0°F$ and actively loses heat energy when the ambient temperature is too high and actively generates heat when the ambient temperature is too low. Heat loss is accomplished by radiation, evaporation, conduction and convection, as we discussed above. Heat generation, on the other hand, is accomplished through chemical reactions, stimulated by muscle activity. Because muscle contraction is so inefficient heat is always generated as a by-product, consequently one of the first reactions a person has to a falling ambient temperature is for the muscles surrounding hair follicles to contract, producing *goose bumps*. Another reaction to low temperature is shivering, resulting from involuntary contractions that cause muscles to quiver, starting deep in the trunk and ending with chattering teeth.[14]

Understanding the mechanisms of thermoregulation in the human body has been one of the success stories of medicine. If the body's core temperature changes by only a few degrees for very short times, an individual is compromised and could die. Therefore medicine focuses on an individual's average temperature to determine that the control mechanisms do not stray outside physiologically pre-established limits. On the other hand, little attention has been given to how much the temperature varies from moment to moment, as long as these random fluctuations are within historically determined margins of variation. Certain periodic changes such as circadian rhythmic changes in temperature, or the variation in the average temperature while awake as opposed to sleeping, have been examined, but these do not concern us here.

Later I examine the statistical fluctuations in body temperature of a normal healthy individual in an effort to determine if these fluctuations contain information about the underlying thermoregulatory mechanisms. Based on traditional wisdom these fluctuations should be random and completely independent of the average temperature of the body. In the next chapter I examine the truth of this "wisdom."

2.3. Respiration regulation

One way to achieve the thermal control just discussed is through mechanical regulation of the ambient temperature, such as with the thermostat in the home. In a similar way, a ventilator, whose function it is to substitute for a patient's own breathing in order to maintain sufficient oxygen levels in the blood, regulates breathing. Andreas Wesele Vesalius developed the theory behind the operation of a ventilator in 1543[15]:

> But that life may be restored to the animal, opening must be attempted in the trunk of the trachea, in which a tube or reed or cane should be put; you will then blow into this, so that the lung may rise again and the animal take in air…. And as I do this, and take care that the lung is inflated in intervals, the motion of the heart and arteries does not stop…

Vesalius also gave the first demonstration in animals showing that life could be sustained by inflation of the lungs with a bellows inserted into the mouth while the chest cavity was opened. This demonstration was not repeated until 1667, 125 years later, when the English physicist, Robert Hooke, an enemy and rival of Isaac Newton, became interested in the phenomenon.[16] But it was not until the late 19th century that mechanical breathing devices became medically acceptable.

There are a number of reasons why a ventilator would be used to assist breathing. These include serious infection, excess fluid in the lungs, respiratory muscle fatigue or a neurological condition that suppresses respiratory drive. Ventilation is a therapeutic technique to maintain adequate levels of oxygen, carbon dioxide and pH while simultaneously minimizing risk and discomfort to the patient. One of the driving forces for the development of ventilators in the early and middle 20th century was poliomyelitis, a virus that causes partial, as well as, complete paralysis. I can recall the *March of Dimes* advertising for donations in movie houses when I was a child. There would be a short film, usually splitting the Saturday afternoon double feature, focusing on someone in a large metal cylinder. This cylinder was an Iron Lung, a device whose origins date back to the 1800s, but which Dr. Phillip Drinker of Harvard University perfected in 1929. The Iron Lung was based on the, then recently discovered, principle of negative thoracic pressure to inflate the chest cavity. A vacuum is created inside the chamber, by an enormous piston at the bottom of the Lung; the negative pressure causes the human lungs to expand and atmospheric pressure pushes air into them.

The person in the film, often a child, would be supine in the iron cylinder within which negative pressures could be applied at regular short time intervals. The head extended out from one end with a rubber seal tight around the neck. The theatre audience saw the face reflected in a tilted mirror above the child's head. She, it was usually a girl in the film, asked the audience to put money into a basket that was being passed around the theatre. Just like a church collection. This was the *March of Dimes*, a name that might sound quaint today, but remember that the cost of the ticket to the Saturday afternoon double feature was only five cents in 1949.

In the 5 years from 1951 through 1955 the annual average number of polio cases in the United States was 37,864 per year. The number of cases dropped by 98.5% to 570 per year during the 5-year interval, 1961 through 1965. This precipitous drop in the number of polio cases was due directly to the introduction of the killed virus vaccine invented by Jonas Salk in 1956 and later the live attenuated virus vaccine by Sabine in 1962. One difference between the two vaccines is that a person cannot contract polio from the killed virus in the Salk vaccine, but a certain small percentage of people do contract polio from the live virus with the Sabine vaccine. The vaccine of Sabine is not unlike that for smallpox with regard to the potential for people contracting the illness. Another difference is that the Salk vaccine is administered by injection into the arm, which was how I first received it; whereas the Sabine vaccine is given orally, which was how my children received it.

I had the good fortune to meet Jonas Salk at the institute the *March of Dimes* built for him in La Jolla, California. He attended a lecture I gave concerning the application of nonlinear dynamics to biomedical phenomena, on the campus of the University of California, San Diego (across the street from the Salk Institute), in the early 1980s. He came up after the seminar, introduced himself and said that we had a great deal to talk about. Indeed, we talked almost daily for the next 10 years, either in his home, my home, at his office, or over the telephone. We became good friends and he strongly influenced the general perspective I have about medicine today.

The worldwide scourge of polio provided the medical pressure for the development of ventilators, but it was WWII and the airplane that first motivated the technological advances necessary for the development of modern mechanical ventilators. There was a direct relationship between how high a bomber would fly and the advances made in artificial ventilation. One could not make use of an aircraft's design to fly above anti-aircraft explosions unless the pilot had a breathing system that could keep him from losing consciousness at high altitudes. This technology developed for pilots' breathing laid the foundation for today's mechanical ventilators.[17]

The Iron Lung was strongly criticized because it caused a venous blood pool in the abdomen and reduced the patient's cardiac output.[18] The positive pressure ventilator, based on aviation technology, has replaced its ignoble predecessor, using the idea that if pressure is greater outside the lungs than it is inside, then air will necessarily flow into the lungs. The closed-loop positive pressure ventilator uses the preceding control theory ideas to accomplish regulation of breathing. Saxon and Myers developed the first mechanical ventilator in 1953 with a closed-loop control technology. Their system used an infrared gas analyzer to monitor the end-partial pressure of carbon dioxide. This information was then fed back to the input to control the pressure and consequently the flow of air into the lungs. This was all done noninvasively. Over the next 30 years sensors were improved and the number of variables being measured was extended, in part, by means of a small tube inserted into an artery to monitor oxygen and carbon dioxide levels in the blood. However, the monitored variables were only used one at a time for feedback control, even though computers had been introduced so that the feedback signal could be processed prior to being applied to the input. Fleur Tehrani[17] did not achieve multiple-variable control of closed-loop positive pressure ventilators until 1987.

What physicians always look for in physiological variables are patterns and breathing is no exception. It is through the identification and interpretation of such patterns that it is determined whether or not a ventilator is performing satisfactorily. A breathing pattern is usually identified as consisting of tidal volume, respiratory frequency and the ratio of the duration of inspiration and expiration. Tidal volume is the difference in lung capacity between normal inspiration and normal expiration in a breathing cycle. It is obvious as to why this is important, since the normal lung capacity determines the average amount of air that enters the lungs in a breathing cycle. The respiratory frequency is the rate at which you breath and like the tidal volume it determines the total amount of air going into and out of the lungs over a given time interval. In fact, ventilation is the product of respiration rate and tidal volume. The therapist chooses various combinations of the three parameters to achieve different outcomes, depending on the pathology of the patient.

Figure 2.2(a) depicts the time series for the breathing of a normal healthy individual lying supine. The time series consists of the sequence of time intervals for completed successive breaths, while a person lies stretched out, relaxed, doing nothing. The average period

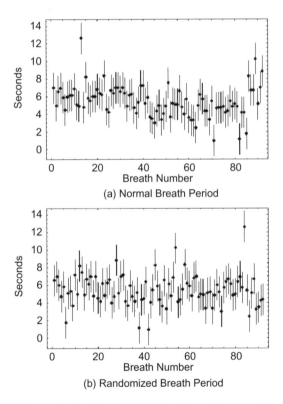

Fig. 2.2: (a) The time series for the period of breathing of a normal healthy individual lying supine are given by the dots. (b) The same data as in (a), but randomized with respect to the breath number. The bars denote one standard deviation above and below the average value.

to complete a breathing cycle is 5.42 seconds; this is the average time from the maximum exhalation of one breath to the maximum exhalation of the next breath. If breathing were a simple periodic process then the time between successive maxima would be the same and the period would be constant. Said differently, the rate of breathing (number of breaths per unit time) would be constant. How much the period or breathing rate changes from breath-to-breath is measured by a statistical quantity called the standard deviation. I shall discuss the meaning of this statistical measure in the next chapter. For the time series shown in Fig. 2.2(a) the standard deviation is 1.75 seconds and is indicated by the vertical bars. It is visually apparent that the variability is quite large and the numbers indicate that the variability in the frequency of respiration is approximately one-third of the average period. What this means is that the respiration frequency can change by as much as one-third from breath to breath in a normal healthy person.

Setting aside the magnitude of the variability for the moment, it is evident that the time series for breathing looks like a sequence of random numbers, going up and down without any apparent pattern. So how is the breath period time series determined to be completely random or not?

There are a number of ways to test a time series for randomness. The most straightforward method is probably to shuffle the data points and determine if the shuffling makes a difference in the properties observed. In Fig. 2.2(b) the data depicted in Fig. 2.2(a) have been shuffled, which is to say, the data points have been randomly moved around with respect to the breath number index, but no data points have been added or subtracted from the data set. Therefore the average value of 5.42 seconds and standard deviation of 1.75 seconds do not change, but the visual appearance of the time series is certainly different from that in Fig. 2.2(a). Even without any analysis, using only the unaided eye one can determine that there is less of a tendency for the data to cluster around certain values in the shuffled data, which is to say, there is less likelihood for successive values of a breath interval to be the same in the randomized case than in the normal case. This difference in the two presentations of the data suggests that breathing is a correlated process, being neither simply periodic nor strictly random. Furthermore, this way of breathing seems to be fairly typical of healthy people.

All the variability shown in Fig. 2.2 is captured in the ventilator parameters with the single number that is the respiratory frequency. The second ventilator parameter is the tidal volume. Figure 2.3 indicates the tidal volume time series for the same individual whose breathing period time series is given in Fig. 2.2. Here again I would expect the time series to yield a single number if the tidal volume were the same for each breath. However it is apparent from the time series data in Fig. 2.3(a) that there is a significant variation in the tidal volume from breath to breath; the mean tidal volume is 0.80 L with a standard deviation of 0.27 L. Again the variability in the tidal volume, as measured by the standard deviation, is approximately one-third that of the average value.

In Fig. 2.3(b) the tidal volume time series data from Fig. 2.3(a) is randomized with respect to breath number. Comparing the renditions of the two time series it is evident that there is a pattern in the normal tidal volume data that disappears when the data is randomized. Here again the data is neither constant, as it would be for a periodic process, nor is it without structure, as it would be for a completely random process.

It is clear from the data set shown in Fig. 2.3(a), typical of normal healthy people, that characterizing breathing with a single respiration frequency and/or tidal volume, ignores patterns within the data. Patterns contain information about the process being examined, and by ignoring such patterns, information is discarded that may be of value to the physician. In fact it is quite possible that the information contained in the variability pattern may in some cases be more important than the average values used in the settings of the ventilator parameters.

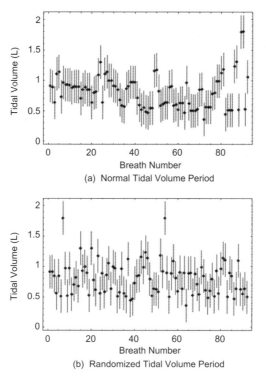

Fig. 2.3: (a) The time series for the tidal volume in liters of a normal healthy individual lying supine are given by the dots and the bars indicate plus and minus the standard deviation. (b) The same data as in (a), but randomized with respect to the breath number. Therefore characterizing the tidal volume by a single number, such as an average tidal volume, for the ventilator, appears to be inadequate.

2.4. Cardiac regulation

I have so far considered how the body regulates its temperature, how breathing can be externally controlled and now I turn the discussion to how the body regulates the beating heart. The human heart is not much bigger than a man's fist, weighs about half a pound and is probably the most efficient pump nature ever designed. The

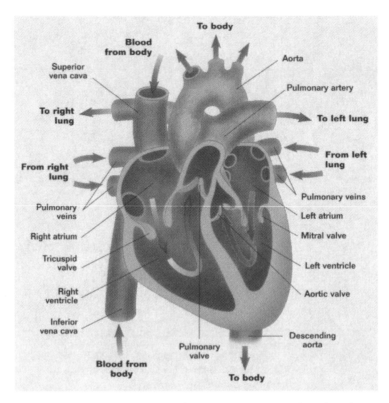

Fig. 2.4: A cross-sectional schematic of the human heart. (Taken from http://www.sirinet.net/~jgjohnso/heartstructure.html.)

heart's function is that of a pump to circulate blood around the body; a remarkable observation first made by Harvey circa 1616. The heart is a hollow sheath of muscle divided into four chambers, as shown in Fig. 2.4. The upper right chamber, called the right atrium, and the lower right chamber, called the right ventricle, draw in oxygen-depleted blood from the body and pump it to the lungs. The upper left chamber, called the left atrium, and the lower left chamber, called the left ventricle, draw in oxygen-rich blood from the lungs and pump it around the body. In this design the atria collect the blood and fill the ventricles, and the ventricles do the pumping. Blood is supplied back to the heart by the coronary arteries.

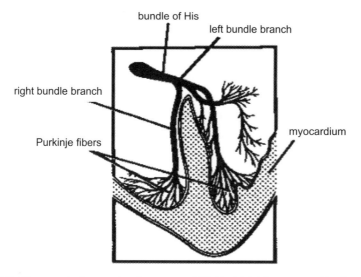

Fig. 2.5: The ventricular conduction system (His-Purkinje) has a repetitive branching structure from larger to smaller scales. The branching structure has been hypothesized to be of a fractal form.

The contractions of the heart are determined, at least in part, by an electrical conduction system that interpenetrates the ventricles and is named after the two physiologists, His and Pukinje, (see Fig. 2.5). Under physiologic conditions, the normal pacemaker cells of the heart are located in the sino atrial (SA) node — a collection of cells with spontaneous automaticity located in the right atrium. The impulse from the SA node spreads through the atrial muscle (triggering atrial contraction) at a speed of nearly one meter per second, so the speed of depolarization takes on the order of 80 ms to cover the atrium. This is called the P wave in an electrocardiogram (ECG). The depolarization wave then spreads through the atrioventricular (AV) node (junction) down the His-Purkinje system into the ventricles. The electrical excitation of the ventricle muscle is referred to as the QRS complex in the electrocardiogram. Finally, there is a T wave signaling the repolarization, that is, the returning to a resting state of the ventricular muscle.

The traditional view of the pumping process is that the AV node functions during normal sinus rhythm as a passive conduit for impulses originating in the SA node, and the intrinsic automaticity of the AV node are suppressed during sinus rhythm. Both the SA and AV nodes are made up of pacemaker cells that would spontaneously fire on their own, with the intrinsic frequency of the pacemaker cell in the SA node being higher than those in the AV node. This view assumes that the AV node does not actively generate impulses or otherwise influence the SA node, but is enslaved by the faster pulsing of the SA node. The ventricular conduction system for the cardiac pulse is shown in Fig. 2.5, where the pulse enters the bundle of His, splits into two pulses of equal amplitude at the two pulses for one, all of equal amplitude. In this manner, a single pulse entering the proximal point in the His-Purkinje network with N distal branches, will generate N pulses at the interface of the conduction network and myocardium. The effect of the finely branching network is to subtly decorrelate the individual pulses that superpose to form the QRS-complex in the ECG shown in Fig. 2.6.

A very different perspective of cardiac conduction was offered by the electrical engineer van der Pol in 1926, a picture that was not universally accepted by the medical community. He and his colleague van der Mark developed a mathematical model in which the beating of the heart is modeled as a dynamic nonlinear oscillator. In such a description the AV node is not simply a passive resistive element in the cardiac electrical network, but is a nonlinearly coupled dynamical partner to the SA node.[19] Another argument for the active role of the AV node is the clinical observation that, under certain conditions, the sinus and ventricular nodes may become functionally disassociated so that independent atrial and ventricular waves are seen on the ECG (AV disassociation). Further, if the SA node is pharmacologically suppressed, or ablated, then the AV node assumes an active pacemaker role. The nonlinear analysis suggests that the SA and AV nodes function in an active and interactive way, with the faster firing SA node appearing to entrain or enslave the AV node.

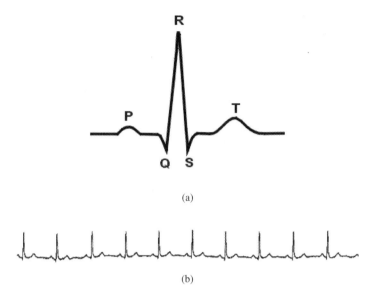

Fig. 2.6: The cardiac cycle. The basic (P-QRS-T) complex depicted in (a) repeats itself again and again as shown in (b).

When pathologies arise that make the intrinsic heart rate too low, or the heart stops beating altogether, it is necessary to introduce an artificial cardiac pacemaker. The first successful implantable cardiac pacemaker was constructed in Buffalo, New York, in 1958 by Wilson Greatbatch. His invention was the result of serendipity and preparation. In 1951, while he was an undergraduate at Cornell University on the GI bill, the phenomenon of heart block was explained to him, during casual conversation on his lunch hour, by two surgeons. In heart block a pulse from the SA node fails to reach the AV node. He understood this problem in terms of a loss of communication between the two nodes and hypothesized a solution to restore communication. The second piece of luck occurred in 1956 while he was working on his Masters degree at the University of Buffalo; he inadvertently put the wrong resistor into a circuit he was constructing using the then newly available transistor. The circuit began to pulsate; each pulse lasted for 1.8 milliseconds and was silent for one second and then the circuit would pulse again. He knew immediately that this was the realization of the theoretical problem he had solved five years earlier; the circuit was mimicing the beating of a heart. The final piece of good fortune was his meeting William C. Chardack, chief of surgery at Buffalo's Veterans Administration Hospital. Dr. Chardack predicted that such an implantable pacemaker would save 10,000 lives per year.[20]

Three weeks after Chardack's motivating prediction, Greatbatch delivered to Chardack's hospital the prototype of an implantable cardiac pacemaker made with two Texas Instrument transistors. In an operating room Chardack and another surgeon, Andrew Gage, exposed the heart of a dog, to which Greatbatch touched the two pacemaker wires. The first cardiac pacemaker took over the beating of the dog's heart, while the three men stood around trying to understand what they were witnessing. This occurred on May 7, 1958. It would be 23 months before Greatbatch would perfect the device to the point where Chardack could implant it into a human being. One of the first of the 10 implants, all of whom had a life expectancy of less than one year, lived for another 30 years.[20]

The original cardiac pacemaker determined the heart rate at a fixed period. However, subsequent generations of artificial cardiac pacemakers assist cardiac function, rather than completely taking it over. Most modern pacemakers wait for a sense event. When it is triggered, the pacemaker produces an electrical pulse of a given width. This small electrical discharge travels down a lead to a bare metal electrode attached to the heart, and stimulates the muscular wall of the targeted heart chamber to contract, thus pumping blood in an organized fashion. Following a pulse, the pacemaker goes into a refractory period for a specified time interval, during which it is unresponsive to cardiac activity. Following the fixed refractory period, the cardiac pacemaker once again monitors cardiac activity. The rate of pulse generation is determined by the programmable pacing rate, length of the refractory period and width of the pulse. The system relies on feedback, in that the cardiac pacemaker contains sensors that enable it to supplement the heart's natural activity. For example, suppose the body's oxygen requirement is not being met during exercise, the pulse frequency of the artificial cardiac pacemaker would then be increased to an appropriate level.

The average heart rate is 60 to 100 beats per minute, but it can occasionally beat as slowly as 40 times per minute while an individual lies in a hammock in the shade from a summer afternoon's sun or strain at 200 beats per minute as the would-be marathon runner pushes herself along the streets of Boston. So the heart rate is certainly not a single fixed quantity, but increases or decreases depending on the load placed on the cardiovascular system. However, no matter what the external stress, the concept of a heart rate is of an average quantity. The public has the notion that if a person is resting, with pleasant music playing in the background, and thoughts of pleasant activities occupying their attention, then the heart rate is essentially the same from minute to minute. This regularity of heart rate is what the medical community has labeled normal sinus rhythm and it is a myth.

In Fig. 2.6(b) we depict the cardiac cycle using the output of an ECG. In this figure the cardiac cycle appears regular, and this apparent regularity contributes to the reason these analogue time series are called normal sinus rhythm by physicians. Measure the distance between peaks, either from P wave to P wave, or between the R peaks in successive QRS complexes, and verify that the interbeat interval is nearly constant from beat to beat. I do this below with a display of the R-R interval time series, just as was done for the inter-breath time interval in Fig. 2.2.

I can look more closely at the properties of the cardiovascular system by explicitly plotting the time interval from one beat to the next in a normal healthy individual. Figure 2.7(a) displays the time interval from one beat to the next in 100 consecutive heart beats, taken from a much longer sequence. I use 100 data points in order to directly compare with the previous data set on breath intervals. It is

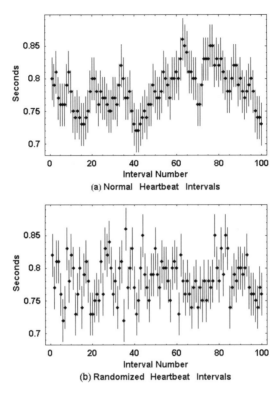

Fig. 2.7: (a) The time series for the interbeat intervals of a normal healthy individual lying supine are given by the dots and the bars indicate plus and minus the standard deviation. (b) The same data as in (a), but randomized with respect to the beat number. Therefore characterizing the beating heart by a single number, such as an average heart rate appears to be inadequate.

evident from the figure that although the heart rate is not constant, with an average time interval of 0.8 sec, which it would be if the beat-to-beat time interval did not change over time, the time interval does not change radically. I shall present more data sets subsequently in order to determine how representative this behavior is. The variation in the average heart rate is approximately 7%, as measured by the standard deviation 0.054 sec, which seems to support the hypothesis of normal sinus rhythm, since $0.054 \ll 0.80$. After all, how much information could be contained in these apparently random changes in the time intervals between beats?

The time series in Fig. 2.7(a) looks like a sequence of random numbers, hopping between values without rhyme or reason. But as in the case of the breathing time intervals we can do a single transformation of the data, that is, randomly shuffle the time intervals, to see if there is any underlying pattern. Shuffling destroys any correlations that exist in the original data set. This test for randomness is depicted in Fig. 2.7(b), where the same average value and standard deviation is obtained as in the original data set because no time intervals have been added or subtracted. The data points have only been randomly moved around relative to one another.

Comparing Fig. 2.7(a) with 2.7(b) we can see that certain patterns of time intervals monotonically increasing or decreasing over short time intervals are suppressed. No such regularity in the time interval data survives the shuffling process in Fig. 2.7(b). Here again we can see from the data shown, which is typical of normal healthy people, that characterizing the cardiac cycle with a single heart rate ignores any underlying patterns of change within the data. As we said earlier, patterns contain information about the process being examined, so by ignoring such patterns in physiologic time series, information is discarded that may be of value to the physician. In fact it is quite possible that the information contained in the variability pattern may be more important than the average values used in settings the parameters in the implanted cardiac pacemaker.

2.5. Averages are not sufficient

The average temperature of the body, the average respiratory frequency, and the average heart rate, all indicate to a worried mother and a concerned physician, the state of a child's health. I know there are systems in the body that regulate body temperature, so if these systems are working correctly then the body temperature stays within a fairly narrow range. If the temperature is outside that range then something is not working right, or there is an infection, and something needs to be repaired. Similarly an overall lethargy or lack of energy may be traced back to a heart that is beating too slowly, or a respiration rate that is working subnormally. When the performance of these systems is below standard, which is to say that the measured indicators deviate from their average values, we are considered ill.

Illness can be diagnosed from a body's regulatory systems working overtime, often indicating the breakdown of one of the many regulatory mechanisms. A fever resulting from the immune system fighting off an infection is an example of the former, and shortness of breath resulting from the failure of the brain to sense the correct oxygen concentration in the blood is an example of the latter. Pathology is manifest by changes in the coarse indicators, which are the average values, of the physiological variables of interest. This is normal physiology. So the average values indicate disease or illness, but are they the best, or most reliable, indicators of health?

It is apparent that the physiologic time series considered in this chapter change erratically with time. This behavior is characteristic of complex phenomena in general and physiologic phenomena in particular. Engineers have been faced with such time series for over a century and they developed ways to model them and during the Second World War, the mathematician Norbert Wiener, synthesized and extended those techniques into what became known as the signal plus noise paradigm.[21] Wiener, as his contribution to the war effort, blended the fields of statistics and communications engineering in order to be able to systematically extract a deterministic message from a statistically fluctuating background. He described complex phenomena, represented by time series, by two distinct mechanisms. One mechanism, if the system is physical, is the deterministic change in time produced by the forces acting on the system described by Newton's law, Bernoulli's law, or whichever physical law is appropriate for the system. The deteministic part of the time series constituted the signal or the message. Of course the message could also be a code containing the kind of information for which Wiener and Shannon originally developed information theory to quantify. The second mechanism is the influence of the <u>environment</u> on the process and this is manifest by rapid irregular variations in the corresponding time series and which Wiener called the noise. It has been referred to as noise ever since.

The appeal of the signal plus noise paradigm in the present context should be evident. The average behavior of the physiologic time series would be the signal; the average heart rate or breathing rate for the examples given in the chapter. The erratic random variations in the time series are produced by the coupling of the physiologic system of interest to the other systems in the body which have very different functions and would be considered noise. This model is fundamentally linear in nature and maintains that the fluctuations do not contain any useful information about the phenomenon of interest. Therefore the time series should be filtered to eliminate this corruption of the information needed by the physician, leaving only the signal on which to base decisions.

This is the model used by the medical community, either explictly or implicitly, in the decision making process. The assumption of linearity, in which the signal and noise are simply added together, is what allows the practitioners to accept the notion that the fluctuations contain no useful information. However every physician knows that linearity is at best an approximation to physiologic phenomenon, since it is evident that the response is almost never proportional to the stimulus. At worst the nonlinearity in complex physiologic phenomena are so important that the fluctuations contain valuable information and filtering the time series discards the very thing the physician needs to accurately diagnose the patient.

What of the variation above and below the average values and the patterns within such variability suggested in the time series? I propose to examine the suggested patterns and seek simple ways of displaying the data to provide information to the healthcare provider. This display of information is not unlike the speedometer that enables the

[Handwritten note: Creating Conclusions of perceived patterns from alterations + Time + environment leads to limited conclusions]

driver to make correct decisions regarding the operation of his/her vehicle. Here the display of the information may enable the healthcare provider to make the best decisions regarding the care of the individual. What I shall argue is that variability, correctly displayed, provides significantly more information regarding a person's health than do averages using the signal plus noise paradigm. This is the domain of *fractal physiology*.

Chapter 3

Even Uncertainty has Laws

Most gamblers are firm believers that good fortune exists, that luck has patterns and that some people are luckier than others. Gamblers know that luck comes in streaks, both good and bad, so when it is good, they bet, and when it is bad, they pass; somewhat like the stockbroker's mantra of "buy low, sell high." Gamblers "know" that luck can attach itself to a chair or card table, to certain dice at the craps table, and that some individuals are blessed with good fortune, while others never get a break. But not just gamblers, ordinary citizens believe in luck in one form or another. This visceral belief is the reason that every newspaper carries a few paragraphs on astrology; it is why the lottery is so popular and why Las Vegas has continued to grow over the past 50 years, through good economic times and bad. Unfortunately this belief in luck may also be aiding and abetting the flowering of superstition that has all but suffocated the popularization of science in the United States and elsewhere. Science celebrates the rational, putting the understanding and control of our complex world at the forefront of human activity. Superstition, on the other hand, places understanding beyond our reach and puts control into the hands of forces we cannot master, surrendering intellect to the caprice of chance.

As in most situations involving understanding of the complexity of the world, everyone is a little bit right and everyone is a little bit wrong and only a few see the forest for the rest the trees block their view. It is the same with knowledge and uncertainty in the information age. As with the previous chapters I will start with a little history in order to retain perspective on contemporary problems and avoid being overpowered by the sheer intricacy of today's society. Furthermore, knowing the roots of a complex phenomenon can sometimes provide guidance in the application of methods of control and insights into what is otherwise obscure. Finally, history enables us to avoid repeating the errors of our intellectual forbearers by anticipating what is to come.

Previously I examined the notion of an average and how that concept has come to dominate the understanding of complex medical phenomena in an industrial society. In particular, I suggested how averages have shaped our notions of health and medicine, much like the parts of an engine are constrained to lie within certain well-defined tolerance intervals. I now need to explain the relationship between the average value of a quantity and the variability of that quantity, independently of the average value. I intend to show that the fluctuating behavior of variables are very different depending on whether they are part of a simple or a complex phenomenon. Of importance to the discussion is the notion that physical systems, upon which most concepts of variability were initially constructed, are basically simple. Consequently the familiar measures of variability developed in the physical sciences are not as useful when applied to the life sciences, because the typical phenomena found in biology and medicine are complex.

But variability, fluctuations and randomness are not the whole story, not by any means. It is true that randomness in the physical sciences took on an attractive form, in that fluctuations were found to follow certain well-defined laws, so that although we cannot predict the survival of an individual, we can predict what will happen to a hospital full of patients over a year's time. However there is also determinism, the underlying forces that project a certain predictability that is embedded within the randomness. This may, in fact, be one workable definition of complexity, the blending of regularity and randomness. The trick is to determine how much of each is in the mix for a given phenomenon and devise methods to exploit that knowledge. You may recall that the signal plus noise paradigm was constructed to do this by assuming the phenomenon to be linear.

Therefore it is necessary to distinguish between a kind of randomness, called chaos, which is generated by the nonlinear dynamical property of a system and the randomness that has historically been called noise. A nonlinear dynamical system is one described by mathematical equations detailing the time changes of the system variables and in which the system's response is not proportional to the applied stimulus. Consequently, the output of a nonlinear system does not have a simple relation to the input. Consider, for example, the temper tantrum of a small child to a mild reprimand. The response is completely disproportionate to the rebuke. Another example is the erratic vibration a car may generate above a certain speed. I had an old Ford with this problem, the chaotic, speed-dependent, vibration turned out to be related to the suspension. Not something I would have suspected given the car's symptoms.

Chaos and noise look very similar, yielding almost indistinguishable time series for the observables of complex phenomena, but they have totally different origins. Noise has nothing to do with the system being investigated and is a property of the environment in which the system is in contact. This is the traditional engineering situation of the time series being signal plus noise, the slow smooth part of the time series is the "signal" and the fast erratic part is the "noise." When the signal-plus-noise paradigm is appropriate it would be correct to smooth the data to eliminate the fluctuations and thereby isolate the signal generated by the system of interest. If, on the other hand, the fluctuations in the time series are generated by chaos, then we most certainly do not want to smooth the data. Chaos is generated by nonlinear dynamical interactions and contains information about the

we are considering our inputs are linear

system, so smoothing or filtering the time series might well eliminate the very thing we want to know. Information theory may provide one of the tools necessary to discriminate between these two sources of variability.

Distinguishing between chaos and noise leads to a discussion of how the traditional modeling in the physical sciences falls short of the needs of the medical sciences. I find it useful to replace a number of the traditional truths arising out of the physics tradition with non-traditional truths that are necessary for the understanding of the medical and social sciences. For example, the emphasis on the quantitative in physics is extended to include the qualitative, and the latter can be quite important in characterizing complex phenomena. Another replacement is the extension of analytic functions used to describe behavior in physical systems to include fractal functions, which are ubiquitous in physiologic phenomena. In the limited context of medicine this replacement describes the transition from traditional physiology to fractal physiology, a new way to understand medicine and how the human body operates. I introduce a new paradigm involving a high wire walker to replace the engineer's signal plus noise in order to incorporate the new truths into a single perspective.

3.1. Randomness and measurement

I briefly trace the development of the concepts of determinism and randomness in science, which is to say in physics, since that is where these ideas were first given quantitative form. The ideas of determinism and randomness developed independently and it is only recently that the historical differences between the two have been seen to be more apparent than real. Some ideas come from 17th century gamblers and mathematicians; some concepts come from 18th century astronomers, who were interested in experimental errors, and some notions come from 19th century physicists interested in the flow of gases and the collision of particles. However, starting from these venerable concepts the mathematicians and physicists of the 20th century, interested in complexity, have provided an array of new ideas. The two different views of science, determinism and randomness, were able to develop, complementing each other, rather than contradicting each other. But the time has come to sort things out and determine what differences are real and what differences are only apparent.

Human beings formulate hypotheses through rational thought, but these hypotheses can only be tested by experiment. Few things are more elegant than a well-reasoned argument from a false premise where no counter argument can break the chain of reasoning. This is apparent to anyone who has ever watched a political debate. Many scientific theories are of this form; the aether that presumably permeated all of space in the 19th century, was such a theory. Logic alone is never sufficient to test a scientific hypothesis, or overthrow a scientific theory. This limitation of logic is why one finds numerous scientific theories to explain a given phenomenon when experimental data is sparse. The data from experiments provide facts that are interpreted to support or contradict a stated hypothesis. Note that a hypothesis can never be proven by experiment, only verified or falsified. However, after a sufficient number of experiments yield data in support of a given hypothesis (verification), and here what constitutes "sufficient" is strictly subjective, the hypothesis is judged to be true. When a hypothesis has reached this level of experimental support, scientists often use the hypothesis to formulate a natural law. Ideally, a natural law encapsulates the results of a vast number of experiments into a simple mathematical expression, or so it has been since the time of Galileo (1564–1642) and his successor Newton (1642–1727).

Once scientists have established a natural law, they no longer make hypotheses; instead they calculate, make predictions and use experiments to test their predictions. If there is a discrepancy between a predicted outcome, based on a natural law, and an experimental result, scientists tend to blame the experiment and/or the experimenter. After all, the weight of experimental evidence is on the side of natural law and prediction. But, if after a sufficiently large number of these new experimental results are obtained, or a significant number of independent verifications of the discrepancies have been made, and the discrepancy between prediction and experiment persists, scientists reluctantly modify the natural law.

One example of such a modification of a natural law was made to Robert Boyle's (1627–1691) law of gases, an elementary physical law that finds many applications in medicine. If we have a certain container with a given volume, containing a gas at a given pressure, and we reduce the volume by a factor of two, then we double the pressure of the gas. The experimental facts are summarized in Boyle's law stating that the product of the pressure and the volume remains constant regardless of the changes made. This law was, of course, the outcome of a vast number of experiments conducted over many years, involving many different gases. Subsequent to the formulation of his law, Boyle continued to do experiments on gases. Some of the results he and other scientists obtained depended on the temperature of the gas before and after the changes in volume or pressure were made, a clear violation of his law. These controversial experimental results led to the formulation of a new natural law, the *perfect gas law*, in which the product of pressure and volume are proportional to the temperature. Thus, Boyle's law has only a limited range of validity; it applies only when the changes in the pressure and volume of the gases do not produce a change in the temperature. This process of testing, summarizing in a law, and testing further, is how science progresses. All natural laws have only a limited range of validity and it is probably a combination of arrogance and fear of the unknown that seduces us into labeling such experimental summaries as "natural laws."

The epitome of natural law was given by Newton's second law of motion and his concept of force. He envisioned that the only reason for the change in the *status quo* was through the application of force. This was summarized in his three laws of motion that are generally memorized for tests in freshman physics classes. The first law is actually due to Galileo and states that all mass has inertia, that is, there is a tendency for all bodies to resist any change in their state of motion. This insight into the nature of the world answered the question raised by the Greeks as to why a javelin continues to move in the air after it leaves the hand of the thrower. To the Greeks this motion was a mystery, because they believed that rest was the natural state of all matter. Consequently, for the javelin to continue moving after leaving the thrower's hand a continuous force must be applied to it at each point along its path. But what could be the source of such a force once the hand of the thrower opens and contact is lost? Inertia recognized that uniform motion, not rest, is the natural state of matter. Therefore, the javelin continues with the same horizontal speed that it had when it left the thrower's grasp, until it strikes the ground. Thus, the first law destroyed the fundamental view of the world handed down from the Greeks.

Newton's second law follows directly from the first because a force must be impressed on a body to change its state of motion. The javelin moves forward at a nearly constant speed, but it also falls to the ground, making a parabolic trajectory. It falls due to the force of gravity, also introduced by Newton. Not that Newton introduced gravity; rather he introduced the idea that gravity is the force that massive bodies exert on one another. Consequently, according to the first law the javelin would travel in a straight line forever, were it not for the forces of air resistance and gravity. This quantitative notion of force was also quite different from the Greek idea of force.

Finally, the third law is the one which almost everyone can quote, but is also the most misunderstood: every action has an equal and opposite reaction. The misunderstanding has to do with where the action and reaction take place. Your weight, due to gravity, pushes down on the chair on which you are sitting. That is the action. The chair, in turn, pushes back on you with an equal and opposite force that just balances your weight. That is the reaction. The two forces act on different bodies, the action force acts on the chair, the reaction force, acts on the sitter. The confusion arises when the action and reaction concepts are applied to the same body; for example, a batter slams the ball over the left field wall for a home run. The bat hitting the ball is the action, but the ball going over the wall is *not* the reaction. The reaction is the ball hitting the bat with equal and opposite force as the bat hitting the ball. But I don't want to delve too deeply into these subtle questions of mechanics, I will only get confused.

From these three laws and the universal law of gravitational attraction, Newton was able to calculate the positions of the planets and their moons through time, unifying the mechanical laws of heaven and earth. Recall that at the time of Newton, most long-range commerce was done on ships, so that anything contributing to the art of navigation influenced the lives of all civilized people. Because Newton's method of modeling the physical world was so successful at predicting the positions of the planets, explaining the motions of the tides and the properties of light, the notions of natural philosophy (science) and prediction became intertwined in the public mind, as well as in the minds of natural philosophers (scientists). It would be a couple of hundred years before natural philosophy would be transformed into science, as we know it today. The idea of expressing a natural law in mathematical form, describing a phenomenon as a cause-effect sequence, and predicting the future from the present, became the scientific paradigm. This was all done using a deterministic set of equations of motion to describe the causal unfolding of phenomena. In this context deterministic simply means that the present determines the future, just as the present has been determined by the past.

Significant differences often occurred between the outcome of experiment and the formulated expectation based on the prediction of natural law. No matter how carefully one prepared the initial state of a system, or how carefully the experiment was controlled, the final measurement always varied from experiment to experiment, leaving the scientist in the position of explaining why the result was not just the single number predicted. Natural laws determined that there should be one outcome resulting from a given experiment starting from a given initial state, so when more than one outcome was observed from a sequence of identically prepared experiments, the notion of error was introduced. The perjorative term error signifies the interpretation of the deviations of the experimental results from the predicted result. This connection between data (experiment) and theory (prediction) was not always appreciated.

Carl Friedrich Gauss (1777–1855) formulated what came to be known as <u>The Law of Frequency of Errors</u> or just the law of errors, using the notion that the average of a large number of experimental results is the best representation of the experiment. Gauss was born 50 years after Newton died. He was a genuine prodigy, doing his father's sums for him when Carl was 3 years old. At the tender age of 32 he published *Theoria motus corporum coelestium*, a seminal work on the application of mathematics to astronomy. In the part of this treatise of interest to us, Gauss analyzed experimental errors in the observation of planetary orbits and their moons. He was particularly concerned with the effects of errors of observation on predictions of future positions of celestial bodies, and in his analysis he laid the foundation for what we now know as statistics and the propagation of error. When his book was published, Gauss had been using his methods of statistical analysis for a number of years, in fact, since he was 18 years old. Note that Gauss' work on "the normal law of errors" was a full 100 years before statistics became a recognized mathematical discipline.

Gauss determined that the errors from experiment, errors being defined as the deviation of an experimental observation from the average value of all such observations, form a bell-shaped curve centered on the average value, as shown in Fig. 3.1. The bell-shaped curve is universal and quantified the notion that small errors occur more often than do large errors. The curve in Fig. 3.1 determines the relative amount of error of a given size in a collection of experimental data. Close to the average value, near zero in the figure, the curve

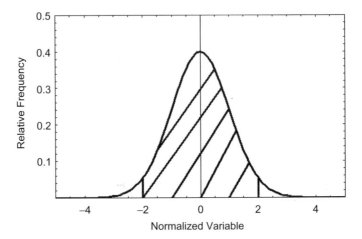

Fig. 3.1: The bell-shape curve of Gauss concerns errors, consequently the experimental value has the average value subtracted so the curve is centered on zero, and the new variable is divided by the standard deviation (width of the distribution) of the old variable and the new variable is therefore dimensionless. The cross-hatched region, between two standard deviations above and below the average value, contains 95% of the errors (the deviations from the average value).

has its maximum value indicating that a large fraction of the experiments yield results in this region. The central maximum not only means that the average value is the most likely outcome of any given experiment, but also that the average value is not a guaranteed outcome of any given experiment. For large errors, far from the average value at the peak of the distribution, near the value of plus or minus two standard deviations,[1] the height of the curve is nearly a factor of 20 smaller than it is at the center. This sharp reduction in the fraction of outcomes indicates that errors of this size are many times less frequent than are those near zero. In fact 68% of all measurements fall between ±1 standard deviations, and 95% of all measuremetns are contained between ±2 standard deviations, the shaded region in the figure. Gauss developed this view of measurement because he

---

[1] The standard deviation is a statistical measure of the width of the bell curve. A very narrow distribution has a small standard deviation implying that the average is a good representation of the data, conversely a large width implies the average is a poor representation of the data.

understood that there ought to be a right answer to the questions asked by experiment. In other words, the outcome of an experiment, even though it has error, is not arbitrary, but rather the outcome follows a predetermined path described by the equations of motion established by Newton.

The height of the Gauss curve determines the relative frequency of the measured deviation of a given size from the average value occurring in a given experiment, when we look at the outcomes of a great many experiments. The average value has the greatest number of outcomes, so that the average value is the most likely measurement. Gauss argued that the physical observable, the quantity over which we actually have control in an experiment, is the average value of the variable and the fluctuations (deviations from the average) are errors, because they are unwanted and uncontrollable. It is the average value that we find in natural laws, such as in the perfect gas law and Boyle's law, and the fluctuations are the annoying consequence of the large number of particles in the system. However, the effect of errors can be systematically eliminated from data by appropriate smoothing methods. In this way science turns attention away from the actual results of experiment and focuses instead on the smoothed data that are interpreted as the average value. The differences among the individual experiments are lost through the process of smoothing; the remaining properties are those of the collective and these averaged properties are those acknowledged by scientists to be important.

It is rarely emphasized that natural laws, like Boyle's law and the perfect gas law, are only true on average. Natural law does not represent the outcome of any particular experiment, but rather it is the property of an ensemble (a collection) of experiments. The outcome of any given experiment deviates from this prediction, being either a little bit more or a little bit less, but on average the law is true. This is the character of all natural laws; they are fundamentally statistical in nature. This is science, a way of describing reality and by its nature it is limited to the domain of observations and therefore to the realm of the uncertain. No matter how certain the theoretical equations, or how reliable the calculations of the prediction, the individual measurements are always a little different from one another. Unfortunately this simple but profound truth is often lost in the scientific cirricula of most universities.

At the end of the 19th century the separation of phenomena into simple and complex was relatively straightforward. A simple system was one that could be described by one or a few variables, and whose equations of motion could be given, for example, by Newton's laws. In such systems the initial conditions are specified and the final state is calculated (predicted) by solving the equations of motion. The predicted behavior of the system, typically a particle trajectory, is then compared with the result of experiment and if the two agree, within a pre-established degree of accuracy, the conclusion is that the simple model provides a faithful description of the phenomenon. Thus, simple physical systems have simple descriptions.

As more particles are added to the system there are more and more interactions, and the relative importance of any single interaction diminishes proportionately. There comes a point at which the properties of the system are no longer determined by the individual particle trajectories, but only by the averages over a large number of such trajectories. This is how the statistical picture of physical phenomenon replaces the individual particle description. The single-particle trajectory is replaced with a distribution function that describes an ensemble of single-particle trajectories and the single-particle equations of motion are replaced with an equation of motion for the probability distribution function. Consider the image of a crowd following the streets in any major city. The individual may move a little to the left or to the right, s/he may walk a little faster or slower, but the flow of humanity is determined by the average movement along the street which sweeps up the individual. Looking out the window of a 20 story building reveals only the flow and not the individual differences. This view from above is what the distribution provides.

In this way the deterministic prediction of a definite future of a single particle, that is characteristic of simple phenomena, is replaced with a predicted collection of possible futures for an ensemble of particles, that is characteristic of "complex" phenomena. The law of errors, the bell-shaped curve, is actually the solution to the simplest of these equations of evolution for a probability density. But Gauss did not know this at the time he wrote down the law of errors, or if he did, he kept it to himself.

Take note that it was not Gauss who first determined the form of the bell-shaped curve; it was the mathematical refugee de Moivré, working in Slaughter's Café, who discovered (first constructed) it. De Moivré's argument entered the 20th century in the hands of Lord Rayleigh, who described the steps of a random walker. The biostatistician Karl Pearson in an inquiry to the magazine *Nature* first articulated the random walk problem in 1905, under the title "The Problem of the Random Walker." Pearson asked for the distribution of steps for a walker who takes a step, then rotates through a random angle, takes another step of the same size, then rotates through another random angle, and continues this process of stepping and randomly rotating indefinitely. This process also became known as the drunkard's walk, the assumption being that a person who is sufficiently drunk will rotate through a random angle after each step to keep from falling down. In the same issue of *Nature*, Lord Rayleigh pointed out that he had solved the random walk question while working on a problem in acoustics. His solution, which he gave, was, of course, a version of the bell-shaped curve first developed by de Moivre and today credited to Gauss. But since his solution contained two variables, rather than one, in this form it is called the Rayleigh distribution. We do love to honor our leaders, scientific and otherwise.

Looking more closely at the properties of the law of errors, consider the sport of archery, which in Gauss' lifetime was much more than a sport. An arrow shot at a target does not miss the bull's eye capriciously. The arrow misses because there is a slight difference in the aiming of the archer from one shot to the next, or there is a mid-course change in the wind speed or direction, or the lighting changes or any of a dozen other factors fluctuate that can influence the aim, release and trajectory of the arrow. However, given the initial angle of the bow and the amount of tension in the bowstring, the speed of the wind, and so on, I could, in principle, predict where the arrow will strike the target. Newton's laws would have no difficulty in determining the prediction, given all the necessary initial conditions. However all these conditions change from instant to instant, so the archer's arrow does not always find the bull's eye. Some strike near the center circle, fewer arrows strike far away, depending on the archer's skill, and some do find the mark. Given a sufficient number of arrows, the distribution of the locations of the arrows striking the target will follow the bell-shaped curve of Gauss; actually because the target is two-dimensional, the curve will that of Rayleigh. In later generations the same reasoning determined the theoretical distribution of hits on the target for a large number of gunshots and the patterns of bombs dropped from planes over Europe during WWII.

Phenomena therefore have a machine-like predictability, even of those that we could say are unpredictable. The 18th and 19th centuries witnessed the introduction and growth of the machine view of the physical world, with the idea that there is a properly determined future that is completely set by the present. When the scientist's prediction is wrong it is because s/he lacks sufficient information about the initial state of the phenomenon in which s/he is interested. In this way, even randomness, in the sense of incomplete knowledge, became subservient to natural law.

In the 19th century, the distribution of Gauss was determined to have myriad applications outside the physical sciences. Near the middle of that century, Adolphe Quetelet (1796–1874) helped lay the foundation of "social physics," which was modeled after celestial mechanics. Quetelet used the law of errors to explain the deviations from the notion of the "average man," not just as a philosophical abstraction, but as a biological reality, being the end result of genetics and evolution. He interpreted human variability as error, much as Gauss had in his theory of physical measurement. This was the beginning of medicine's misunderstanding of complexity.

Francis Galton (1822–1911), the cousin of Charles Darwin (1809–1882), extended the application of Quetelet's ideas and championed the use of the law of errors in the social sciences and made the following observation concerning the emergence of patterns in large data sets:

> I know of scarcely anything so apt to impress the imagination as the wonderful form of cosmic order expressed by the "Law of Frequency of Error." The law would have been personified by the Greeks and deified, if they had known of it. It reigns with serenity and in complete self-effacement, amidst the wildest confusion. The huger the mob, and the greater the apparent anarchy, the more perfect is its sway. It is the supreme law of Unreason. Whenever a large sample of chaotic elements are taken in hand and marshaled in the order of their magnitude, an unexpected and most beautiful form of regularity proves to have been latent all along.

Galton is here marveling at the fact that even though the individuals in a population vary, the characteristics of the population as a whole are themselves stable. The fact that statistical stability emerges out of individual variability has the appearance of order emanating from chaos. It is this observation of the orderliness of error that eventually led to the assumption that the regularity of the average is more important than the variability of the individual.

Galton constructed a device to show how universal is the law of error. The device consisted of a box-like object with a piece of glass on one side. Inside the box was a hopper at the top with a single hole that was large enough to release one marble or ball bearing at a time, (see Fig. 3.2). The falling marble encounters a number of rows, made up of equally spaced pegs, with spaces between the pegs large enough for the marble to fall between two of them. The positions of the pegs in successive rows were offset so that a marble falling through a space in one row would encounter a peg in the next row, just below the hole. In this way a marble would fall from the hopper

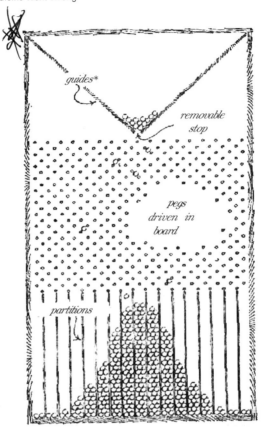

Fig. 3.2: Galton's pegboard, the *Quincunx*, was used in the 19th century to show by experiment the form of the bell-shaped distribution of the Law of Frequency of Errors.

at the top, strike the peg beneath and the distance traveled would be such that the marble would bounce either to the right or to the left by no more than one space with equal frequency. Suppose it bounced to the right, it would then fall through the space between the center peg and the first peg on the right. Just below this hole is centered another peg. It would strike this peg and the marble would then bounce to the left or to the right, with equal frequency. At each row the marble would bounce one space to the right or to the left, until it eventually reached the last row where there are partitions and it would fall and stack up at the bottom of the *Quincunx* Galton's board (see Fig. 3.2).

The random walk process mentioned earlier and the trajectory of the marble falling within Galton's board are clearly equivalent. Consider the vertical direction in the Galton board to be time and the horizontal direction to be a one-dimensional path along which the random walker moves, so that the step the drunk takes is to the right or the left depending on which way the marble bounces. But making this choice of stepping to the right or to the left is like tossing a coin; since I cannot know in advance which way a marble will bounce at any given peg. The time to take a step in the random walk is the time it takes the marble to fall from one row to the next, so if the rows are equally spaced, the time it takes the walker to take a step is also constant. Most of the time there are nearly as many bounces (steps) to the right as to the left, so the central part of the board has the most marbles. The decrease in the number of marbles as we proceed from the center outward has the characteristic fall-off of the bell-shaped curve. In fact this is an experimental realization of Gauss' curve.

The law of errors took on an even greater virtue through the quiet development in physics of the *central limit theorem* during the 19th century and which blossomed in mathematics in the early part of the 20th century. The central limit theorem gained in importance because, it was unrestricted in its application and is equally true in discussing populations of cows, people, atoms or stars. When taken together any collection has properties that are independent of the properties of the individuals making up the collection. The central limit theorem is true when a sufficiently large number of independent random variables, with finite variance, all having the same statistics, are added together to yield a random variable with a bell-shaped statistical distribution. This gave the weight of mathematical certainty to the application of the bell-shaped curve to all manner of phenomena in the physical and life sciences.

The mathematical statement of the central limit theorem is a generalization of the experimentally motivated law of errors. But do not get the impression that the application of the bell-shaped curve was universally accepted without criticism; the French mathematician and theoretical physicist Poincaré (1854–1912) made the observation that[22]:

> All the world believes it firmly, because the mathematicians imagine that is a fact of observation and the observers that it is a theorem of mathematics.

This is, of course, the worst recommendation a scientist can make about a theory. It is not unlike the dancers in a Broadway show who say a person is a good singer, and the singers who say the same person is a good dancer. The person is always good using someone else's criteria, never one's own. In the case of the law of errors, Poincaré was alluding to the fact that a given phenomenon does not necessarily satisfy the conditions of the mathematical theorem in a given sequence of measurements. We shall find that Poincaré's prophetic words of caution anticipated the true nature of complexity, that in the characterization of the uncertainty of a complex phenomenon, the bell-shaped curve proves to be too simple, regardless of how often it is used today.

Let us make this discussion a bit more human by considering the notion of luck a bit more carefully. Luck is one of those superstitions that has a great deal of support, particularly among individuals that take an intuitive approach to life. Very often the examples people cite in support of their luck arise from misunderstandings of the nature of probability. Suppose you were asked to make the following wager in order to pass the time in a long line waiting for the doors to open to your favorite movie. Your friend turns to you saying; "I bet there are at least two people in this line that have the same birthday." You look around and estimate that there are 50 people in line and thinking that there are 365 days in the year, that it is unlikely that any two people would have the same birthdate. In fact, you think it is a fools bet, particularly since your friend wants to bet even money. You and your friend announce your bet to the twitter of laughter and 53 people give you their birthdays. Sure enough there are two individuals with the same birthdate. You shake your head and pay your friend muttering; "You lucky bastard."

But was it luck?

Let's apply what we know about probabilities to the problem. An individual has a birthday on a given day, let's say September 20. A second individual can have a birthday on any day of the year other than September 20 and there are 364 such days. Therefore the probability that the second person was born on a day other than September 20 is 364/365. Let's suppose the birthday of the second person is July 7. Now consider a third person, the probability that this person's birthday differs from the first two, so it is not September 20 or July 7, is obtained using the remaining 363 days to be 363/365. Consequently, the probability of not having the three people share a birthdate is

$$\frac{364}{365} \cdot \frac{363}{365}.$$

In this way the probability of a group of any size not having the same birthdate can be constructed.

The probability concerning the friendly wager in the movie line can now be constructed. The probability that no two people in the line having the same birthday is obtained by calculating the product of 52 factors

$$\frac{364}{365} \cdot \frac{363}{365} \cdots \frac{312}{365}.$$

Of course, this calculation is done on a computer, yielding 0.025. This means that the probability of any two people in the line having the same birthdate is 0.975, a virtual certainty. Therefore, luck is not always what is appears to be and perhaps your friend knew something that you did not.

3.2. Chaos and determinism

Various statistical views of the physical world were evolving in parallel with the predictable, deterministic, machine worldview of Newton.

But we do not have time to pull together all these various threads and describe how the fabric of the history of science was woven, so we restrict ourselves to certain developments that influenced the understanding of averages and fluctuations. The modern development of the Newtonian view began at the end of the last century with the previously mentioned scientist Poincaré. In his investigation into the stability of the solar system, Poincaré discovered that most of the equations describing real physical phenomena have solutions less like traditional planetary orbits and more like the path of a tornado, being irregular and unpredictable.

The story begins in 1887 when the king of Sweden, Oscar II, offered a prize of 2,500 crowns to anyone who could determine the stability of the solar system. The king wanted to know if the planets in their present orbits would remain in those orbits for eternity, or whether the moon would crash into the earth, or the earth would hurl itself into the sun. This was and still is one of the outstanding questions in astronomy, not so much for its practical importance, since the sun will in all probability cool down before the planets careen off course, but because we ought to be able to answer the question with 19th century physics and it is frustrating that we cannot.

One of the respondents and the eventual winner of this contest was Henri Poincaré. Although Poincaré did not answer the King's question, he won the competition for a number of reasons. The 270-page monograph he submitted in 1890 established a new branch of mathematics that allowed for the geometrical interpretation of equations of motion, particularly for many bodies moving under the mutual Newtonian gravitational attraction. He was able to show that only the two-body problem had periodic solutions in general, that is, if the universe consisted of only the sun and a single planet, then the planetary orbit would periodically return to the same point in space in its orbit around the sun. Furthermore, he was able to show that if a third body was added to the above two-body universe, and if that body was much lighter than the other two, then the orbit of this third body would in general not be periodic, but instead its orbit would appear to be an incredible tangle, much like a plate of spagetti. One hundred years later we find that this theoretical orbit is, in fact, a fractal in space, but to Poincaré the complexity of the three-body problem indicated that the King's question regarding the stability of the solar system could not be answered. It was this analysis that introduced chaos into the scientist's picture of the world, even though the word as applied to dynamics would not be implemented for nearly another century.

Thus, the conclusions of Poincaré's analysis, since they contradicted the clockwork universe of Newton, were, for the most part, ignored for nearly 100 years by the scientific community. His conclusions only re-emerged in the early 1960s through the seminal work of the meteorologist Ed Lorenz. Perhaps the results of Lorenz were more acceptable because weather is a physical phenomenon that most people believe ought to be unpredictable due to its complexity, not like celestial mechanics that ought to be predictable because of its apparent simplicity. Whatever the reasons, scientists slowly began to accept the new view of the importance of nonlinear interactions and subsequently the ubiquity of chaos in determining the properties of complex phenomena. The unpredictability of the nonlinear world would not remain secret, no matter how long scientists averted their eyes.

Empirically, there developed two parallel, but distinct, views of the physical world. One world was deterministic, in which phenomena are predictable and without error. The second world is stochastic, with uncertain predictions confined to a range of possibilities. The boundary between these two worldviews began to fade in the 20th century when scientists began to interconnect the descriptions of particles observed with a microscope with phenomena seen with the unaided eye. What scientists found is that the principle of reductionism, which had served so well in the building of machines, did not seem to work so well when biological systems were investigated. In the reductionistic view, natural law is applicable to the smallest scales of a system and from the dynamics at these smallest scales the behavior at the largest scales of a system can be determined. In fact, large-scale dynamics should be the consequence of the small-scale dynamics plus the notion of additivity. The paradigm for this perspective was the application of Newton's equations for individual particles to the interactions among celestial bodies, the latter planets being the addition of a huge number of the former particles.

Knowledge about the world depends on the paradigm of science; from the nurse to the neurosurgeon, from the biologist to the businessman, from the physicist to the fakir, all have been taught to believe in the knowability of the world. Furthermore, this knowability is a consequence of the world being an orderly place. Newton's laws epitomize the orderliness of the world; laws that imply cause-effect sequences, synthesizing rational thought and empirical fact. The notions of cause and effect go like this: given a definite constellation of wholly material things, there will always follow from it the same observable event. If the initial configuration is repeated, the same event will follow. I flip the switch and the light goes on, I turn the key and the car starts, I step on the pedal and the car accelerates. If the expected outcome does not occur, the light refuses to go on, the car remains quite and does not move, then I am sure that the constellation from which I started was not the same. There is something wrong, and this *wrongness* is material and has injected a material difference into the initial constellation, which has always worked before. The present determines (is the cause of) the future. This concept of cause has been elevated to center stage in science. Of course, I could say a great deal more about causality, in the manner of Aristotle, but that would not serve the purpose of the present discussion.

No predictive model of the world has to date served any better than reductionism. This logical view has taken society from the Renaissance, through the industrial revolution, into the information age of the computer. Ironically, it is with the aid of the computer that people have been able to determine that the linear cause-effect sequence imposed through idealized natural law does not, in fact, exist in the real world, or at least not in the way it was once thought to exist. Simple predictions only work if investigators do not look too closely at their results. Wherever and whenever scientists carefully examine nature they see deviations from predictions and are forced to conclude that things are more complex than they had thought. This may be another possible definition of complexity. A system is complex when it is sufficiently intricate that the familiar notion of cause and effect is no longer applicable, which is to say, that the "effect" is no longer entailed by the "cause" in any readily identifiable way. But what is science without causality?

One example of the loss of causality is contained in a single pacemaker cell of the human heart. This cell will spontaneously fire without the application of an external stimulus. In a human heart large numbers of these cells fire together, acting in concert, to produce heart beats. The cardiac control system acts to ensure that the interval between beats is physiologically reasonable. What is the cause and what is the effect in such a system? This phenomenon is cyclic, so that seen as a whole, the cycle has no beginning and no end. In mathematics this is called a limit cycle and like a roller coaster the system travels through the limit cycle, periodically returning to every point, but undergoing possibly complicated motion between returns. This picture suggests that complex pheomena may not be described as a strict sequence of cause-effect relations. If we must abandon the traditional notion of causality isn't the worldview given over to anarchy?

The answer to this last question is no. It is a false choice. What I have done, in part, is to confuse natural law with certainty, when, in fact, they are not the same thing. I have argued that laws of chance (Bernoulli's law of large numbers and Gauss' Law of Errors) can be just as rigorous as natural laws. In fact, from one perspective, what has made the laws of nature so successful is that they are the accumulations of laws of chance, and they are extremely good approximations to the situation where the laws of chance combine to give overwhelming likelihood. An example of this overwhelming likelihood may help to clarify what I mean. Consider Newton's laws which refer to the motion of individual particles and yet they are applied to planets and suns, each with billions upon billions of such particles. Surely the application of these laws in celestial mechanics is a vindication of the laws of chance, with the individuality of distinct particle trajectories being washed out in support of collective behavior. This was the beginning of reductionistic theory in which the overall behavior of the collective could be deduced from the superposition of the actions of the fundamental constituents. In fact, the laws of the microscopic (small scale) variables were guessed at from observations of the macroscopic (large scale) variables.

Look at this question of complexity from a different and perhaps deeper perspective and see what conclusions can be drawn. Historically scientists assumed that complex phenomena eluding causal description could be broken down into smaller and smaller pieces, eventually reaching some smallest elements. These smallest elements are assumed to follow cause-effect interactive chains described by deterministic natural laws, such as Newton's laws of motion for the constitutive particles of a planet. In the case of a planet's motion the system is simple and the microscopic laws aggregate to give a macroscopic law sometimes of the same form and sometimes not. This argument is similarly applied to complex phenomena. Scientists then argue that the loss of causality is only apparent and given sufficient knowledge about the initial states of the microscopic elements we could predict the unfolding of even very complex phenomena with absolute certainty. Into this aesthetically pleasing picture we now introduce the mechanism of chaos.

Chaos is a technical term from dynamics that has gained popular coinage. It is a short-hand for the difficulties associated with interpreting the typical solutions to nonlinear dynamical equations. Chaos is important because every physical interaction is nonlinear and so too are the dynamical responses to disturbances; consequently, the solutions to these equations of motion are chaotic. This means that even at the smallest scales, instead of simple cause and effect, uncertainty is deterministically generated. This uncertainty is amplified in aggregations of such dynamically unstable microscopic elements that make up macroscopic phenomena. Thus, statistical laws only describe what scientists assume ought to be causal, those being phenomena that are sufficiently simple that microscopic differences are washed out in the macroscopic description.

Dynamics, the branch of physics that first gave us confidence in natural law by describing the planets in their orbits around the sun, has, since Poincaré, undermined that confidence by showing that planetary orbits are the exception in science and not the rule. There are a number of conclusions that have been drawn from these observations. First, complex phenomena are nonlinear in nature, for if such phenomena were linear, the principle of superposition would apply and the cause-effect chain of reductionism would prevail. Second, the nonlinear interactions between the fundamental elements of a system yield chaotic behavior in the dynamics of those elements. Such microscopic chaos means erratic fluctuations, like the movement of the gas molecules in the room, only more complicated. More complicated because in addition to the randomness there can also be long-range correlations introduced by corresponding long-range correlations.

Given the microscopic randomness produced by chaotic solutions to nonlinear deterministic equations of motion there is a corresponding loss of certainty about the past, present and future states of a system. Without an exact specification of the present and future state there can be no direct causal link established between the two. Even in the case of the beating human heart, where there is a self-generating cyclic behavior in the pacemaker cells, normal sinus rhythm no longer implies a regular heartbeat. Instead there is a distribution of heartbeats produced by the nonlinear dynamical nature of the cardiac control system. Thus, even by segmenting the cardiac dynamical cycle into local elements, say the cardiac conduction system between the SA and AV nodes, the implicit uncertainty of the dynamics prevents the exact prediction of when the next pulse will be generated. This assumption regarding the uncertainty of pulse generation is tested subsequently.

Eventually there is the loss of the philosophical underpinnings of reductionism in modern science because deterministic natural law has been shown to be statistical regardless of the historical path followed. Statistics wash out causality, undercuts the naïve expectation of exact predictability, and sweeps away reductionism as well. It is not that the dynamics at one level (scale) of observation is independent of the dynamical laws operating at the next lower level (scale) of observation, but rather the laws at the lower level (scale) do not uniquely determine the dynamics at the higher level (scale). Consequently it is worthwhile to investigative how nonlinear dynamics modifies some of the notions of complexity in science.

3.3. Wisdom is not static

The conventional wisdom of the physical sciences has been collected into a small number of traditional truths, which, through repetition, have become so obvious that they are all but impossible to call into question. These truths are associated with the traditional description scientists use to explain natural phenomena. I briefly review those truths here with a view towards exposing their weaknesses as revealed by scientists during the last half of the 20th century. I present a traditional truth and then as counterpoint I present a corresponding non-traditional truth that has developed over the last 20 years or so. I do this for four traditional truths and their non-traditional counterparts to explain a new vision of science, one that is more amenable to the life sciences. I first began this exercise for understanding the shift in thinking about nonlinearity and complexity in science in the early 1980s[23] and since then my perspective has undergone a number of fundamental changes.[24]

The first traditional truth, if I ask any physical scientist, is that scientific theories are quantitative and this is not a matter of choice, but of necessity. Lord Rutherford's view is fairly typical of the physical scientist. His view of science was dismissive of nearly everything that was not physics as evidenced by his comment:

> All science is either physics or stamp collecting. Qualitative is nothing but poor quantitative.

This was a criticism of the tradition in such disciplines as developmental biology, clinical psychology and medicine, where large amount of qualitative and quantitative data were gathered, but no quantitative predictions were made due to the absence of quantifiable theories. By the end of the 19th century there were two main groups in natural philosophy: those that followed the physics of Descartes and those that followed the dictates of Newton[25]:

> Descartes, with his vortices, his hooked atoms, and the like, explained everything and calculated nothing. Newton, with the inverse square law of gravitation, calculated everything and explained nothing. History has endorsed Newton and relegated the Cartesian constructions to the domain of curious speculation (p. 763).

Thus, the first traditional truth is: *If it is not quantitative, it is not scientific.*

This truth is a visceral belief that has molded the science of the 20th century, in particular, those emerging disciplines relating to life science and sociology have by and large accepted the need for quantitative measures. It would be counterproductive to say that the quantitative perspective is wrong; instead, I suggest that it is overly restrictive. Quantitative measurement does not help the physician to understand the aesthetic judgment made in viewing a painting or listening to music, and through such understanding, help in reaching decisions concerning the quality of life of a patient. For such considerations the qualitative as well as the quantitative aspects of the world need to be examined.

The first of the non-traditional truths in counterpoint is that scientific theories can be qualitative as well as quantitative. The most venerable proponent of this view in the last century was D'Arcy

Thompson,[26] whose work, in part, motivated the development of catastrophe theory by René Thom.[25] Thompson's, and later Thom's, interest in biological morphogenesis stimulated a new way of thinking about change; not the smooth, continuous quantitative change familiar in many physical phenomena, but the abrupt, discontinuous, qualitative change, familiar from the experience of "getting a joke," "having an insight" and the bursting of a bubble (most recently this might be the *e.com* bubble). This abrupt qualitative change in the dynamics of a system is called a bifurcation.

One example of a mathematical discipline whose application in science emphasizes the qualitative over the quantitative is the bifurcation behavior of certain nonlinear dynamical equations. A bifurcation is a qualitative change in the solution to an equation of motion obtained through the variation of a control parameter. Suppose we had a mathematical model of the respiratory system. In such a model system the amount of oxygen in the blood might be a control parameter, so that when the oxygen density decreases the respiration rate is stimulated to increase, thereby bringing the oxygen density back to the desired level. In a nonlinear periodic system of unit period, changing the control parameter might generate a sequence of bifurcations in which the period of the solution doubles, doubles again, and so on producing more and more rapid oscillations with increasing changes in the control parameter. In such systems the solution eventually becomes irregular in time and the system is said to have made a transition from regular rhythmic oscillations to chaotic motion. This may represent, for example, the apparently disorganized gasping for breath a runner experiences after pushing the envelope too far in a race. These kinds of transitions are observed in fluid flow going from laminar to turbulent behavior. The áperiodic behavior of the final state in such bifurcating systems has suggested a new paradigm for the unpredictable behavior of complex systems. This successive bifrucation path is only one of the many routes to chaos.

Another route to chaos is the onset of intermittency through the change in a control parameter of the same kind we discussed in homeostatic control theory, except that the system must be non-linear. In this intermittent transition to chaos a nonlinear system's motion is regular (periodic) when the control parameter is below a critical value. When the control parameter is above this critical value the system has long intervals of time in which its behavior is periodic, but this apparently regular behavior is intermittently interrupted by a finite-duration burst of activity, during which the dynamics are qualitatively different. The time intervals between bursts are apparently random. As the control parameter is increased above the critical value, the bursts become more and more frequent, until finally only the bursts remain. A mathematical model with this type of intermittent transitions was used by Freeman[27] to describe the onset of low-dimensional chaos experimentally observed in the brain during epileptic seizures.

These arguments suggest the form of the first non-traditional truth: *Qualitative descriptions of phenomena can be as important, if not more important, than quantitative descriptions in science.*

The second fundamental truth has to do with how scientists mathematically model phenomena. The scientists that developed the discipline to describe the motion of celestial bodies accepted that mechanics and physics are described by smooth, continuous and unique functions, often referred to collectively as analytic functions. This belief permeated physics because the evolution of physical processes are modeled by systems of dynamical equations and the solutions to such equations were thought to be continuous and to have finite slopes almost everywhere. Consequently understanding comes from prediction and description not from the identification of teleological causation, that is, science is concerned with the *how* of phenomena not with the *why*. Galileo condensed natural philosophy into the three tenants:

Description is the pursuit of science, not causation.
Science should follow mathematical, that is, deductive reasoning.
First principle comes from experiment, not the intellect.

Thus, the stage was set for Newton who was born the year Galileo died. Newton embraced the philosophy of Galileo and in so doing inferred mathematical premises from experiments rather than from physical hypotheses. This way of proceeding from the data to the model is no less viable today than it was over 300 years ago and the second tradition truth emerges as: *Physical observables are represented by analytic functions.*

Of course, there are a number of phenomena that cannot be described by analytic functions. For example, ice melts in your soda and water freezes on your driveway, rain evaporates after a spring shower and water condenses on the outside of your glass in the summer sun. Laminar fluid flow becomes turbulent, as experienced by the veteran airline passenger or the rafter traveling down the Colorado River. In these phenomena, scaling is found to be crucial in the descriptions of the underlying processes, and analytic functions are not very useful, for example, in describing the lightning shown in Fig. 3.3(a) or the neronal cell in Fig. 3.3(b). Implicit in the idea of an analytic function is the notion of a fundamental scale that determines the variability of the phenomenon being described by a function. When examined on a relatively large scale the function may vary in a complicated way. However, when viewed on a scale smaller than the fundamental scale, an analytic function is smooth and relatively unvarying.

All complex phenomena contain scales that are tied together in time, space or both. The idea that a phenomenon has a characteristic scale is lost, and most things, like the structures shown in Fig. 3.3, have multiple scales contributing to them. This is the situation for a function said to have a scaling property. Note that here, the function is used as a noun, not a verb, and refers to a mathematical representation of data that can be traced on a graph. If such a function scales, the structure observed at one scale is repeated on all subsequent scales, the structure cascades downward and upward, never becoming smooth, always revealing more and more structure. Thus, we have structure, within structure, within structure. As we magnify

Fig. 3.3: (a) The ramified network of electrical discharge in lightning, leading to a fractal network in space is obvious. (b) Cell body located in the cellebellum of a hedgehog, obtained from http://faculty.washington.edu/chudler/cellcerebell.html.

such functions there is no limiting smallest scale size and therefore there is no scale at which the variations in the function subside. Of course this is only true mathematically; a physical or biological phenomenon would always have a smallest scale. However, if this cascade of interactions covers sufficiently many scales, it is useful to treat the natural function as if it shared all the properties of the mathematical function, including the lack of a fundamental scale. Such phenomena are called fractal, and they are described by fractal (non-analytic) functions.

Everything we see, smell, taste, and otherwise experience are in a continual process of change. If we experienced such changes linearly, meaning that our response is in direct proportion to the change, we would be continuously reacting in response to a changing environment, rather than acting in a way to modify that environment. Such behavior would leave us completely at the mercy of outside influences. But the changes in the world are not experienced linearly. We know from experiments conducted in the 19th century, on the sensation of sound and touch that people do not respond to the absolute level of stimulation, but rather they respond to the percentage change in stimulation. These and other similar experiments became the basis of a branch of experimental psychology called *psychophysics*. These experiments supported a mathematical postulate made much earlier by Daniel Bernoulli in 1738 involving utility functions, which were intended to characterize an individual's social well being.

Suppose we both invest money in the stock market. I invest in a stock selling at $100 per share and buy 10 shares. You invest in a stock selling at $10 per share and buy 10 shares. Now suppose that at the end of the month my stock is valued at $101 per share and yours is valued at $11 per share. We both make $10 profit on our relative investments, but who is happier? You, who made a 10% profit or I, who made a 1% profit? The utility function of Bernoulli specifies that individuals respond to percentage changes, not to absolute changes. So accordingly you are happier than I am even though we both made the same amount of money, because your percentage change was greater than mine. This form of the utility function captures an observed aspect of these biological and social phenomena; they appear to be independent of scale size. So even though particular utility functions are no longer fashionable in the social sciences, the notion of scale-independence seems to be outside such parochial considerations.

When an erratic time series lacks a fundamental scale, it is called a statistical or random fractal. This concept has been used to gain insight into the fractal nature of a number of complex physiological systems: the chaotic heart, in which the cardiac control system determining the intervals between successive beats is found to have coupling across a broad range of dynamical scales; fractal respiration, in which breathing intervals are not equally spaced, but are basically intermittent in nature, is also seen to have a mixture of deep regular breaths with short irregular gasps; and for the first time scientists have seen that the fluctuations in human gait contain information about the motorcontrol system guiding our every step.

Thus, the second non-traditional truth is: *Many phenomena are singular (fractal) in character and cannot be represented by analytic functions.*

The third traditional truth has to do with prediction. The time evolution of physical systems are determined by systems of deterministic nonlinear dynamical equations, so the initial state of the system completely determines the solution to the equations of motion and correspondingly, the final state of that physical system. Consequently the final state is predicted from a given initial state using the dynamic equations. Until recently, this view prompted scientists to arrogantly assert that the change of these observables in time, are absolutely predictable in the strict tradition of the mechanical clockwork universe.

The third traditional truth is: *The dynamics of physical systems are predictable by means of their equations of motion.* This mechanistic view of the universe, with its certainty about the future, probably more than any other single thing, has made people uncomfortable with science. In large part this discomfort arises because such certainty about what is to come is inconsistent with most people's experience of the world.

It is true that the Newtonian model provides a valid description of the motion of material bodies. But do his laws imply absolute predictability? The particular results with which we are concerned are the áperiodic (irregular) solutions to his equations of motion, which are a manifestation of the deterministic randomness (chaos) that can arise from nonlinear interactions. Since chaotic processes are irregular, they have limited predictability and call into question whether phenomena are quantitative, analytic and predictable. Indeed a number of scientists have proposed that the statistical properties observed in complex phenomena do not arise from the large number of variables in the system, such as I argued earlier, but arise instead from the intrinsic uncertainty associated with chaotic dynamics.

The third non-traditional truth is related to the fact that most phenomena are described by nonlinear equations of motion and consequently their paths of motion from the past into the future are chaotic. Chaos implies that the phenomenon of interest is fundamentally unstable. Recall the example of the archer used earlier. I argued that it was a slight change in the angle of the bow, a small increase in the tension of the bowstring, a modest deviation in the direction or speed of the wind, that could divert the arrow from the bulls eye and therefore give rise to the bell-shaped distribution. What if the smallest change in the angle of the arrow caused it to completely miss the target, or a puff of wind turned the arrow around in mid-flight? Is it absurd would such behavior be completely unstable. Truth be told, that is in fact the nature of chaotic behavior; an unexpected dynamics that often runs counter to intuition. We do not often see such behavior in the physical world, at least not on the time scale we live our lives. But don't forget that earthquakes, volcano eruptions and tornadoes, are natural phenomena that are large scale, unpredictable and devastating. At an earlier time it would have been argued that our inability to predict these things was a consequence of incomplete knowledge and given sufficient information we could make such predictions. However, given the chaotic solutions to complex dynamical systems, we now know that predicting the time of occurences of earthquakes, volcano eruptions and tornadoes is beyond our present day science.

Less dramatic phenomena come to us from the biological and social realms, which are filled with examples of such instabilities. In fact the best selling book, *The Tipping Point*,[28] was about what I would call a critical point, separating two qualitatively distinct kinds of behavior, connected by a bifurcation.

Phase space is the geometrical space in which a dynamical system lives. It consists of coordinate axes defined by the independent variables for the system and is useful for specifying a system's dynamics. Each point in such a space corresponds to a particular set of values of the dynamical variables that uniquely defines the state of the system. The point denoting the system moves about in this space as the phenomenon evolves and leaves a trail that is indexed by the time. This trail is referred to as the orbit or trajectory of the system and each initial state produces a different trajectory. It is often the case that no matter where the orbits are initiated they end up on the same geometrical structure in phase space. This structure to which all the trajectories are drawn as time increases is called an *attractor*. An attractor is the geometrical limiting set of points in phase space to which all the trajectories are drawn and upon which they eventually find themselves. After an initial transient period, that depends on the initial state, an orbit blends with the attractor and eventually loses its identity. Attractors come in many shapes and sizes, but they all have the property of occupying a finite volume of phase space.

A simple example of a nonlinear dynamical system is the one that initiated the new Renessance in nonlinear dynamics, the Lorenz system. Ed Lorenz was a meteorologist at MIT when he discovered the set of nonlinear equations involving three variables, which now bear his name. He discovered these equations, according to his excellently readable book, *The Essence of Chaos*,[29] in a discussion with a colleague Barry Saltzman at the Traverlers Weather Center. He had been looking for equations that would generate non-periodic solutions, since he had discovered such behavior a year earlier. The way in which he discovered chaos in these equations is quite resonable, as he describes in his book:

> At one point I decided to repeat some of the computations in order to examine what was happening in greater detail. I stopped the computer, typed in a line of numbers that it had printed out a while earlier, and set it running again. I went down the hall for a cup of coffee and returned after about an hour, during which time the computer had simulated about two months of weather. The numbers being printed were nothing like the old ones. ...Instead of a sudden break, I found that the new values at first repeated the old ones, but soon afterward differed by one and then several units in the last decimal place, and then began to differ in the next to the last place and then in the place before that. In fact, the differences more or less steadily doubled in size every four days or so, until all resemblance with the original output disappeared somewhere in the second month. This was enough to tell me what had happened: the numbers that I had typed in were not the exact original numbers, but were the rounded-off values that had appeared in the original printed output. The initial rounded-off errors were the culprits; they were steadily amplifying until they dominated the solution. In today's terminology, there was chaos.

The equations that Lorenz adopted, modeled a particular physical phenomenon involving the heating of a fluid from below and followed the flow of the fluid in a vortex-like motion. What is of interest here is the notion of an attractor and the idea of bifurcations. Figure 3.4 depicts the solutions to the Lorenz model on a plane and the third variable is not shown. An attractor is shown on top left for a specific set of values of the control parameters in the model. The system is initiated in the vicinity of the origin and the trajectory winds its way over to the right, where it is drawn onto an attractor. Once on the attractor, the trajectory stays there forever. What makes this significant is that all trajectories in this space will be drawn to this attractor, independently of their initial states. On the upper right is depicted how one of these variables changes with time. It is evident from this time series that after an initial transient period the solution

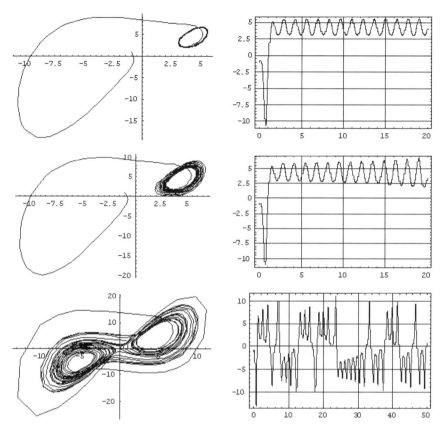

Fig. 3.4: The solutions for the Lorenz model for three different values of one of the control parameters are depicted. On the left are the attractors in the phase space of two of the three variables. On the right are the corresponding time series solutions for one of those variables.

is periodic; the motion is on an attractor (limit cycle) and consequently is periodic.

I increase one of the control parameters, the coupling between the temperature and the flow field, and obtain the new attractor shown in the middle graph of Fig. 3.4. Here the attractor is seen to have much more structure than before and the corresponding time series on the right is not quite periodic, having an increasing amplitude with time. The qualitative behavior of the attractor has bifurcated, by which is meant that the qualitative behavior of the solution has changed.

Finally, in the lower figure the solution has bifurcated again by an increase in one control parameter beyond what is called a critical value, afterwhich the attractor becomes "strange." The trajectories on such strange attractors are chaotic, as are the time series associated with the individual variables. This is apparent from the áperiodic behavior of the time series on the lower right. Here again any initial state of the system is drawn to the strange attractor over time, after which all memory of the initial state is lost.

Whether or not a dynamical system is chaotic, is determined by how two initially nearby trajectories cover the system's attractor over time. As Poincaré stated, a small change in the initial separation of any two trajectories may produce an enormous change in their final separation (sensitive dependence on initial conditions) — when this occurs, the attractor is said to be "strange," in the sense used by Lorenz. The distance between two nearby orbits increases exponentially with time, which is to say, the distance between trajectories grows faster than any power of the time. But how can such a rapidly growing separation occur on an object, the attractor, which has a finite volume? Don't the diverging orbits have to stop after a while? The answer to this question has to do with the structure necessary for an attractor to be chaotic. The transverse cross section of the layered structure of a strange attractor is fractal and this property of strange attractors relates the geometry of fractals to the dynamics of chaos.

Chaos can be understood by watching a taffy machine slowly pulling taffy apart and swirling it in the air only to push it together again, or observing a baker rolling out dough to make bread. The baker sprinkles flour on his breadboard, slams a tin of dough on the surface and sprinkles more flour on top of the pale ball. He works the rolling pin over the top of the dough, spreading it out until the dough is sufficiently thin; then he reaches out and folds the dough back over onto itself and begins the process of rolling out the dough again. The Russian mathematician Arnold gave a memorable image of this process using the head of cat inscribed in a square of dough, as shown in Fig. 3.5. After the first rolling operation the head is flattened

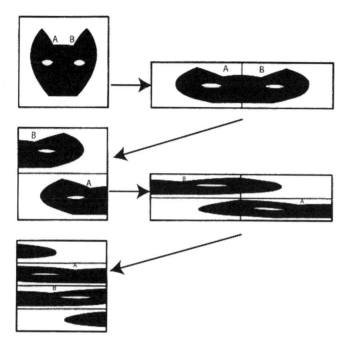

Fig. 3.5: Arnold's cat being decimated the stretching and folding operation that accompanies the dynamics of a chaotic attractor. After only two operations the cat's head is unrecognizable.

and stretched, that is, it becomes half its height and twice its length, as shown in the top right side of the figure. The dough is then cut in the center. The right segment is lifted and placed over the left segment to reform the initial square. This is the mathematical equivalent of the baker's motions. The operation is repeated again and we see that at the bottom, the cat's head is now embedded in four layers of dough. Even after only two of these transformations the cat's head is clearly decimated. After 20 such transformations, the head will be distributed across 2^{20} or approximately one million layers of dough. There is no way to identify the head from this distribution across the layers of dough.

The above argument, originally due to Rössler, turns out to be generic. Two initially nearby orbits, represented by the labels A and B on the cat's head, cannot rapidly separate forever on a finite attractor.

At first the two points are adjacent with A to the left of B. After one iteration B is to the left of A and slighly above. After two iterations A is to the right of B and B is slightly below. The relative positions of the two points is not predictable from one iteration to the next. In general, the attractor structure (cat's head) must afford ample opportunity for trajectories to diverge and follow increasingly different paths (different layers of dough). The finite size of the attractor insures that these diverging trajectories will eventually pass close to one another again, albeit on different layers of the attractor which are not directly accessible. One can visualize these orbits on a chaotic attractor being shuffled by this process, like a dealer in Las Vegas shuffles a deck of cards. This process of stretching and folding creates folds within folds ad infinitum, resulting in the attractor having a fractal structure in phase space. All this is summarized in the third non-traditional truth: *Nonlinear, deterministic equations of motion, whether discrete or continuous, do not necessarily have predictable final states due to the sensitivity of the solutions on initial conditions.*

Finally, the above considerations suggest the fourth traditional truth: *Physical systems can be characterized by fundamental scales such as those of length and time.* Such scales provide the fundamental units in the physical sciences, without which measurements could not be made and quantification would not be possible. In the present context, these scales manifest themselves in the average values of distributions. This is consistent with the bell-shaped distribution of Gauss, where we have a well-defined average value of the variable of interest and a width of the distribution to indicate how well the average characterizes the result of experiment. Observe that a Gaussian distribution is completely characterized in terms of its width and average value, but what of distributions whose average values diverge? Such distributions, for example, inverse power laws, are known far and wide outside the physical sciences. One example of such a distribution was obtained by the Italian sociologist/engineer/economist Vilfredo Pareto[30] to describe the distribution of income in western societies and is discussed further in the next chapter. The distribution of income is fundamentally different from the bell-shaped curve. The inverse power law is seen to have a long tail in Fig. 3.6, which extends far beyond the region where the bell-shaped curve is confined. This long tail often indicates a lack of scale and this lack of scale can be manifest in the divergence of the variance of the dynamical variable.

In the same way that the bell-shaped curve can be related to Newton's laws and simple dynamical systems, so that the curve describes the distribution of errors, the inverse power-law curve can be related back to unstable dynamical systems. Very often the heavy-tailed distribution is a consequence of intermittent chaos in the underlying dynamics, so that large values of the variable occur more often than would have been expected based on the stable dynamics of simple systems. This does not happen to the archer in the real

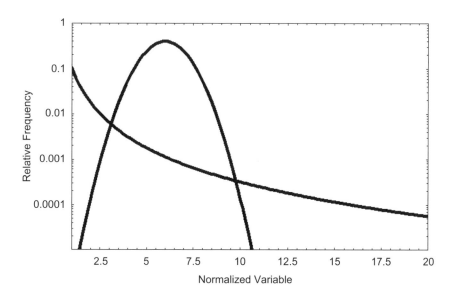

Fig. 3.6: The bell-shaped curve of Gauss is here compared with the inverse power law of Pareto. The much longer tail of the Pareto distribution is evident, so large events are much more likely in the latter case than in the former.

world, because the distribution of his misses is bell-shaped. However, in the much more complex world of business, the forces determining a person's level of income interact in varied and sundry ways to produce an income level over which you and I have little or no control most of the time. In the case of income we might have the illusion of control, but if that were in fact true, then the inverse power-law distribution of income would not be universal, which it appears to be.

The fourth non-traditional truth is: *Natural phenomena do not necessarily have fundamental scales and may be described by scaling relations.* Classical scaling principles are based on the notion that the underlying process is uniform, filling an interval in a smooth, continuous fashion. The new principle is one that can generate richly detailed, heterogeneous, but self-similar structure at all scales. The non-traditional truth is that the structure of many systems are determined by the scale of the measuring instrument, and such things as the length of a curve are a function of the unit of measure, for example, the length of a fractal curve depends on the size of the ruler used to measure it. In Fig. 3.7 is a sketched curve with many twists and turns. It is clear that using a ruler of a given size, as in each sketch, I obtain a specific length of the curve; one which ignores a great deal of small scale structure. In the third sketch, using a ruler of one-quarter the length of the first, the total measured length of the curve increases, but not by a factor of four. The increase is due to the additional small scale structure that with the smaller ruler is seen to contribute to the measured length. The fourth and final sketch shows that by reducing the ruler by a factor of 20 from the first again increases the measured length of the curve. A factal curve is one that does not have a characteristic scale so the size of the ruler can be reduced indefinitely and the measured curve becomes infinitely long. The mathematical way in which the unit of measure and the measured length of the curve are related is called a scaling relation and it is nonlinear for a fractal curve.

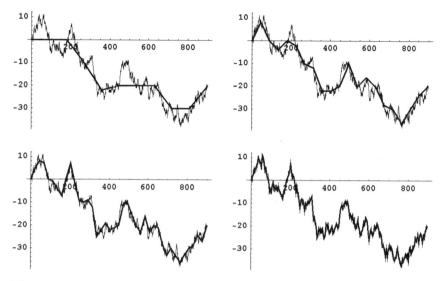

Fig. 3.7: The top left curve is a sketch of a "fractal" curve 1000 units long with a ruler of length 100 units. The second curve indicates a measured length of the curve using a ruler 50 units long. The third curve indicates a measured length of the curve using a ruler 25 units long and the fourth uses a ruler 5 units in length.

3.4. A new tradition

In this chapter, I have argued that science has five basic characteristics. First of all, science is guided by natural law, which is to say that scientists believe that there is order in the universe and this order is summarized in the form of natural laws. Second, science explains events, phenomena, and processes in terms of their being deduced from such natural laws. Here the term explanation is free of any teleological implications; in fact, explanations are restricted to the realm of description. Third, science is testable against an objectively existing empirical world, so that a law remains a law only so long as no contravening body of experimental evidence is found. Consequently, science is not absolute but relative, relative to a vast amount of experimental information and not to the whim of the latest shaman. Fourth, the conclusions of science are tentative, since new experiments may change laws and therefore modify the conclusions drawn from the existence of the particular forms of such laws. Fifth and finally, scientific predictions are falsifiable, and such an ability to falsify again relates back to the experimental evidence obtained from the empirical world.

As I said, complexity is implicit in the system-of-systems concept. Figure 3.8 introduces a measure of complexity that starts at zero, increases to a maximum value and then decreases to zero again. This measure is actually based on a modification of the information idea introduced earlier. Start at the left of the figure with the dynamics of one or a few variables and denote this as being simple, since the equations of motion and their solutions are well known. Complexity increases with an increasing number of variables or degrees of freedom as we proceed up the curve to the right. The mathematics of

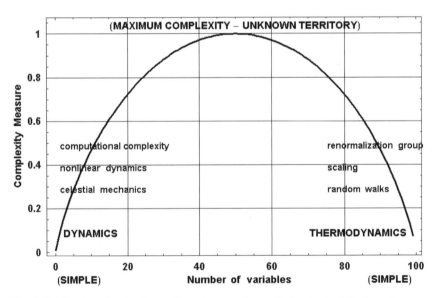

Fig. 3.8: Here is a sketch of a possible measure of complexity. At the left where there are few variables, determinism and the dynamics of individual trajectories dominate. At the right where there are a great many variables, randomness and the dynamics of probability densities dominate. At the center, the peak of the complexity curve, maximum complexity involving both regularity and randomness together, with neither dominating.

such systems includes nonlinear dynamics, control theory and so on, referring to mathematical formalisms that are fairly well understood. There has been a substantial body of mathematical analysis developed regarding complexity and its measures, and the broad range over which mathematical reasoning and modeling have been applied is rather surprising. One class of problems, which defines the limits of applicability of such reasoning, is "computational complexity," the distal discipline on the left side of the complexity curve. A problem is said to be computationally complex if to compute the solution one has to write a very long algorithm, essentially one as long as the solution itself. Applications of this quite formal theory can be found in a variety of areas of applied mathematics, but herein I avoid these more formal issues and focus attention on the influence of nonlinear dynamics in such theories and the subsequent notion of complexity derived from this influence.

The commonality of the techniques on the left side of the complexity curve is they are all methods for handling the deterministic single particle trajectories of the system's dynamics. However, as the complexity continues to increase with an increasing number of variables, the techniques become less useful and blend into what we do not yet know. What becomes evident is that determinism alone is no longer sufficient to understand complex phenomena.

On the other side of Fig. 3.8, where the number of degrees of freedom is very large, we have equilibrium thermodynamics, and the phenomena are again simple. The mathematics describing such systems involve partial differential equations for the evolution of the probability distribution functions, renormalization group theory and scaling of the coupling across scales, all of which are mathematical formalisms that assist our understanding of the physical and life sciences. The mathematics here is designed to handle random phenomena and describe the system's dynamics. Here again ascending the complexity curve, moving from right to left, the system being modeled increases in complexity and the mathematical tools available again become less useful.

The unknown territory lies between these two extremes of simplicity. The area of maximum complexity is where one knows the least mathematically, as well as experimentally. The peak is where neither randomness nor determinism dominates, nonlinearity is everywhere, all interactions are nonlocal and nothing is ever forgotten. Here is where turbulence lurks, where the mysteries of neurophysiology take root, and the secrets of DNA are hidden. All the problems in the physical and life sciences that have confounded the best minds are here waiting in the dark corners for the next mathematical concept to provide some light. It is the failure to acknowledge that medical phenomena are balanced on this peak of complexity that has led to so much misunderstanding in medicine, based on the idea of linear homeostatic control.

I subsequently argued that the four non-traditional truths introduced in this chapter provide a way of looking at medicine that was not available during earlier times. Non-traditional truths must be employed because the complexity of the human body does not lend itself to the traditional truths of science. These non-traditional truths form the basis of *fractal physiology*, a discipline that has been developing over the past two decades. In this new discipline resides the elements of complexity, where the traditional truths are replaced with non-traditional truths. In this new view:

i) quantitative scientific theories are augmented by qualitative ones;
ii) phenomena are not only represented by analytic functions, but by non-analytic functions as well, in order to describe singular and erratic behavior:
iii) the time-paths of phenomena cannot always be predicted by equations of motion, for arbitrarily long times;
iv) natural systems do not necessarily possess a fundamental scale and must instead be described by scaling relations.

To recap: Historically thermodynamics was the first scientific discipline to systematically investigate the order and randomness of complex systems, since it was in this context that the natural tendency of things to become disordered was first quantified and made systematic. As remarked by Schrödinger:

> The non-physicist finds it hard to believe that really the ordinary laws of physics, which he regards as prototypes of inviolable precision, should be based on the statistical tendency of matter to go over into disorder.

In this context the quantitative measure of "disorder" that has proven to be very valuable is entropy, and thermodynamic equilibrium is the state of maximum entropy. Of course, since entropy has been used as a measure of disorder, it can perhaps also be used as a measure of complexity. If living matter is considered to be among the most complex of systems, for example the human brain, then it is useful to understand how the enigmatic state of being alive is related to entropy. Schrödinger maintained that a living organism can only hold off the state of maximum entropy, that being death, by absorbing negative entropy, or *negentropy*, from the environment. He points out that the essential thing in metabolism is that the organism succeed in freeing itself from all the entropy it cannot help producing while alive, much like the wake left by a passing ship. Here we note that negentropy is nothing more than information.

In a thermodynamics context, complexity is associated with disorder, which is to say, with limited knowability, and order is associated with simplicity or absolute knowability. This rather comfortable separation into the complex and the simple, or the knowable and the unknowable, in the natural sciences breaks down outside the restricted domain of thermodynamics. The fundamental ambiguity in the definition of complexity, even adopting entropy as the measure leads to multiple definitions and indeed to inconsistencies between these definitions. A special role is played by the information entropy of Wiener, Shannon and Kolmogorov that relates the physical properties of the system to the concept of information.

A completely random system would have maximum thermodynamic entropy and therefore maximum complexity. This concept of entropy is very useful for studying large-scale, physical and chemical systems at or near equilibrium. On the other hand, an ordered system, such as a living organism or a society, would have a low thermodynamical entropy and would therefore be simple under this classification scheme, in which complexity is proportional to entropy. Since this conclusion runs counter to the expectation that living systems are among the most complex in the universe, the definition of thermodynamical entropy cannot be simply applied to a system in order to determine its complexity. However, I do not want to abandon the notion of entropy altogether since it is tied up with the order-disorder property of a physical system. Thus, I intend to explore some of the extensions and refinements of the entropy/information concept to see if these will serve our needs better.

The information measure of complexity is low on the left in Fig. 3.8 for two reasons. The first is the small number of variables that have deterministic equations and predictable trajectories. The very predictability of the dynamics makes the information content low, along with a corresponding low level of complexity. The second, and related reason, is the generic chaos that exists in even the simplest nonlinear dynamical system. One measure of chaos is the constant (increasing) rate of information generated by the dynamical process, thereby inhibiting any predictions about the system's future. This diminishing ability to predict increases in both the information gained with each measurement and the level of complexity. Thus, as the number of nonlinear interactions increases, so too does the level of chaos and the amount of information being generated, thereby also increasing the level of complexity.

The same information measure is maximum when the system is completely random due to it having an infinite number of interacting variables. The organization in such systems is minimal, however, so that although information generation is great, the level of complexity is low. The measure of complexity that is used in Fig. 3.8 is the product of the information gained and degree of organization (suitably defined) in order to obtain the level of complexity. The fact that the level of organization is zero in an uncorrelated random process suppresses any contribution to complexity that the high entropy level may contribute at the far right. As the randomness becomes correlated and the degree of organization increases, moving from the extreme right to the left, up the curve in Fig. 3.8, the complexity of the phenomena increases.

But enough of the abstract, what can be learned by examining some selected data sets?

Chapter 4

The Uncertainty of Health

The map is not the territory but an admonition that one should not confuse one's perception of the world with the truth. A dean once assured me, as Chair of a Physics Department and new to the academic world, that the perception of truth was, in fact, more important than the truth. However, this cynical view does not serve the working scientist very well. The scientist's vocation is to reduce the differences between the scientific map and the territory, and in so doing, enhance the understanding of the world. Unfortunately, most people do not have a direct experience of the territory, and so their map or mental picture of the world is a collage. People only know little pieces of the territory first-hand and they fill in the large blank spaces with what they have heard, what they have read and assumptions they have made, with which they are comfortable. These assumptions are largely not open to question because people do not know they have made them.

For example, people rarely think about how tools have transformed their map of the world. The invention of the automobile invalidated the view of the man on horseback. The adoption of the telephone undercut the view of the letter writer. Each advance changed the perception of the world in unforeseen ways, culminating in a modern map that is almost totally the result of human construction, with little room for the variability of nature. The purpose of this chapter is to show how this variability ought to be reintroduced into the world view.

Phenomena with complicated and intricate features, having both the characteristics of randomness and order, are considered complex. However, there is no consensus among scientists or poets as to what constitutes a reasonable quantitative measure of complexity. Therefore, any list of complexity traits is arbitrary and idiosyncratic, and mine is as follows:

(i) A complex system typically contains many elements and each element is represented by a time-changing variable.
(ii) A complex system typically contains a large number of relations among its elements. These relations are modeled by a number of interdependent dynamical equations that determine how the system changes over time.
(iii) The relations among the elements are generally nonlinear in nature, often being of a threshold or saturation character, i.e. the strength of the coupling of one variable to another often depends on the magnitude of the variable.
(iv) The relations among the elements of the system are constrained by the environment and often take the form of being externally driven or having a time-dependent coupling. This coupling is a way for the system to probe the environment and adapt its evolution for maximal survival.
(v) A complex system is typically a composite of order and randomness, but with neither being dominant.
(vi) Complex systems often exhibit scaling behavior over a wide range of time and/or length scales, indicating that no one scale, or a few scales, are able to characterize the evolution of the system.

Having written these six properties explicitly, it is worth observing that I do not know how to consistently take these conditions into account in a general dynamical model of complexity. However, no one I know or know about, can construct such a general model either. On the other hand, it is clear that physiologic phenomena satisfy each of these criteria, and so to have a better understanding of how these characteristics can be quantified, I continue with the use of an informative measure of complexity.

An issue related to the information approach to understanding complexity is the principle of reductionism. In a nutshell, this principle states that the process of understanding implies processing data for the purpose of arriving at generalizations. Such generalizations are very efficient descriptions of the world, reducing what is needed to remember and enhancing the ability to communicate with one another. It is much simpler to communicate a law than to communicate the results of thousands of experiments upon which the law is based. In its strong form, reductionism states that to understand complex phenomena, one only needs to understand the microscopic laws governing all the elements of the system that make up the phenomenon. This reasoning implies that once I understand all the parts of a process, I can "add them up" to understand the total. Like Euclid's geometry, the whole is the sum of its parts, but scientists find that such sums are an incomplete description of natural phenomena; the large-scale description that balances the micro-description of reductionism is systems theory. From the latter perspective, a nonlinear system very often organizes itself into patterns that cannot be understood using only the laws governing the dynamics of the individual elements. This self-organization constitutes the emergence of new properties that arise, for example, in phase transitions, when the local weak interactions among the particles in a fluid become the strong long-range interactions of a solid. Living beings, too, cannot be understood using reductionism alone, but require the adoption of a more holistic perspective. This change in perspective, from the reductionistic to the holistic, in some ways resembles the passage from deterministic to probabilistic knowledge. In both cases, the meaning of "knowledge" changes with the changing perspective.

In the reductionistic theory of the physical sciences, there is typically a transformation to a set of variables that can be considered independent, one from the other. In physics, these variables are referred to as modes and they are exact descriptors of linear systems. In non-linear systems which are non-chaotic, these modes can be used to describe the transfer of energy (information) from one mode of the system to another. However, when the system is chaotic, such a modal representation breaks down and so does the reductionistic view. The preference for reductionism is a vestige of the "machine age" having to do with closed systems, consisting of passive, observable elements. This view is no longer appropriate for describing the complex phenomena of the "scaling age" or "information age" in which systems are open, consisting of dynamically interactive and often unobservable elements. This change in perspective, in part, reflects a willingness to step beyond the safe, deterministic models of traditional quantitative science into the mostly unexplored regions of qualitative science.

In this chapter, I examine measures on the mathematical desert whose expanse is strewn with the remains of discontinuous functions, processes that are chaotic and phenomena that are unpredictable. However, there is an oasis where the reader might be able to relax and think about the things learned and ponder why the more traditional ideas, that are crucial in the physical sciences do not apply in the realm of medicine.

At this oasis, I examine some properties of scientists as well as science, as understanding people and the tools they choose to use, often helps us define what they do or hope to do, and why they are successful or not. Becoming comfortable with the people may allow the reader to become equally comfortable with the science they do, perhaps not with the technical knowhow, but certainly with the knowwhy and knowwhat.

I have emphasized average values, the law of errors and the central limit theorem. Now I turn to an examination of the kind of systems and phenomena that can be properly described by the law of errors, but more importantly, I examine some systems that cannot be so described. I do this to convince the reader that the law of errors is not the appropriate description of fluctuations in physiological phenomena. A number of examples of phenomena are provided outside the physical sciences that are described by inverse power-law distributions. These are examples of a ubiquitous distribution in economics, sociology, urban growth and biological evolution, and reference many more examples.

The difference between the bell-shaped curve and the inverse power-law curve depicted in Fig. 3.6 summarizes the fundamental difference between simple physical systems and complex medical phenomena. I hope to overcome the bias of the average value and the bell-shaped curve in this chapter and to indicate what is purchased at this price.

In addition to the arguments given above, there might also exist other reasons of why scientific theories do not provide a satisfactory description of the complex phenomena, given the present state of knowledge. It is not sufficient for science to describe the world within the laboratory; it must also faithfully describe the world in which we live. It seems clear that reductionism is not sufficiently robust to describe systems where the pattern of information flow often plays a more important role than that of microscopic dynamics. However, science still requires that the macroscopic rules be consistent with the microscopic ones. If new properties emerge, even if it is not possible, in practice, to predict them from microscopic dynamics, the new properties must be implicit in the microscopic equations. This is the weak reductionistic assumption and is part of the objectivity principle. A chemical reaction may be too difficult to be explained from a purely quantum mechanical perspective at the present time, but nevertheless, no violation of quantum mechanics is expected to take place during a chemical reaction. The analogous situation arises at a higher level for the biological properties of a cell that cannot be understood in terms of chemistry alone. The understanding of the global properties is achieved from a holistic point of view, but the emerging properties have to be compatible with a weakly reductionistic perspective. Otherwise, one would be tempted to imagine different laws for different space and/or time scales, or different levels of complexity. This, in turn, inhibits any possible mechanistic (objective) view of reality. I stress that in this perspective, the principle of objectivity, namely the objective existence of mechanical laws, does not necessarily mean that the laws are deterministic, but a seed of randomness may be involved. Actually, it was argued elsewhere that a seed of randomness must be involved in any fundamental description of reality.[31]

4.1. The different kinds of scientists

In the previous chapters, I dealt mostly with the nature of science and very little with the characteristics of scientists. It is often overlooked that science is a social activity and even those who choose the legendary solitude of Sir Isaac Newton, act to claim priority, putting in a good word for friends while undermining the aspirations of competitors. However, documenting the all too human frailties of scientists is not my intention; besides, such documentation would require volumes for its completion. A more modest aim is to identify five types of scientific personalities, so as to avoid the solemnity that such pontification so often produces, and so I have chosen to name these types *sleepers*, *keepers*, *creepers*, *leapers* and *reapers*. It is necessary to go over these personality types so that readers may divest themselves of the comfortable fiction that science follows a single well-defined procedure for gaining knowledge and that scientists are all cast from the same mold. I had previously defined the first four types of scientists as[32]:

> *Sleepers* are those scientists that have given up the pursuit of knowledge and mostly pass on to the next generation what they and others have previously learned. They no longer actively participate in the research that advances knowledge, but rather organize, categorize

and make logical the rather haphazard procedure known as scientific research. Consequently, with regard to involvement in research, they are asleep.

Keepers are those researchers that refine, redo and carry out experiments that add to the next significant figure in the fine structure constant or the gravitational constant. On the theoretical side, these investigators establish new proofs or new physical models of well-understood phenomena. As a group, these scientists put forward all the objections to new theories and explain in extraordinary detail why this particular experiment must be wrong or that proof is flawed. These are the maintainers of consistency, searching for the internal contradictions in theories and experimental results; these are the keepers of the *status quo*.

Creepers are the worker bees of the sciences. They are the investigators that take us from what is presently known, to what has been predicted, but have not yet been established. These scientists carry out the systematic investigations that build out and away from our present understanding to what has been guessed, divined or predicted through intuition and/or theory. These are the experimenters and theoreticians that fill in the blank spaces on the landscape at the frontiers of science, the mapmakers that show us how to creep (cautiously make our way) from one oasis to the next.

Leapers are the type of scientists deified in myth. These are the individuals that create the new laws, against all odds and popular scientific belief with the penetrating insights that change the discipline forever, who function as magicians completely outside the circumscribed rules of science. The mathematical physicist Mark Kac, once remarked in talking about such scientists as Sir Isaac Newton and the Noble Laureate Richard Feynman, that they were magicians. He explained that we could do what a genius did if only we were considerably smarter. However, Kac maintained that it was not clear how we could ever do what a magician does. These scientists make connections within and among phenomena in ways that logic cannot follow and that may take the creepers years to verify. These are the scientists who leap to the correct conclusion.

However, let us now add an additional personality type to this list:

> **Reapers** are scientists that seem to acquire, sometimes earned and sometimes not, all the accolades and recognition afforded scientific discovery. Very often, it is not the originator of a scientific theory, nor the first to perform an extraordinary experiment, who is recognized for their original contribution. It is often the last person to discuss the phenomenon who is credited with its understanding. This does not imply any deception on the part of the beneficiary, although that does occasionally happen, but merely points out one of the foibles of being human. We prefer to give the credit for an idea, discovery or scientific breakthrough to a single individual rather than to a group or to a sequence of individuals, and the most recent is usually the most obvious recipient of our attention.

Of course no one fits neatly into these categories all the time. Individual scientists at different stages of his/her career may assume one kind of personality at one time and a different kind of personality at another time. Albert Einstein, perhaps the greatest *leaper* of all times, in his theories of special and general relativity, certainly functioned as a *keeper* with regard to the description of microscopic phenomena and his rejection of the Copenhagen interpretation of quantum mechanics. He also functioned as a *creeper* in his five papers on the nature of physical diffusion. While it is true that his work in 1905 on diffusion is completely original, this work was not the first to accurately describe diffusive phenomenon. A French physicist named Bachelier, a student of Poincaré, published a paper in 1900 on the diffusion of profit in the French stock market. Bachelier developed the same equations and found the same solutions to those equations, as did Einstein five years later. Consequently, one could argue that Einstein was a *reaper* with regard to recognition that should have been more properly given to another. It is not that Einstein sought such recognition; it was thrust on him and consequently, denied to others. What is apparent is that Einstein was never a *sleeper*. He moved with ease and without contradiction from one personality type to the other, depending on the problem he was addressing.

The above five definitions of scientific personalities do not capture the actual meaning of what it is to be in a given group. Consider the example of a sleeper. Mechanics is one of the fundamental areas of physics and its purpose is to describe the motion of the material bodies using Newton's laws. I was taught mechanics by Mr. Johnson (I never knew his first name), a person who would never achieve fame or fortune in science, but who loved to teach the formalism. There was something about capturing the physical world and the way things moved in a maze of symbols, which gave him deep satisfaction. Although he could not extend the formalism that others had developed before him, Mr. Johnson could appreciate their elegance and a sense of detailed wonder that emanated from the pristine beauty of logical necessity. From the equations of motion, to the solution of a specific problem, the understanding of why a racetrack is banked on one side, to knowing how a parachute works, were wonders he shared with us. Mr. Johnson was not charismatic; in fact, one might say he was aloof. He was balding, always in a white shirt, dark tie and blue suit with chalk on his sleeves and he almost never smiled. He loved talking about physics, not to make it simple or even understandable, but to share it with the class (there were eight of us). No critique of Hamlet, no discussion of Socrates, and no dissection of Tennessee Williams ever matched the warmth Mr. Johnson radiated when he discussed the equations of a swinging pendulum or the motion of a spinning top. The sleepers in science are neither irrelevant nor unimportant; they prepare the next generation of scientists and when they do it well, it is like poetry, particularly when it is done with tenderness and affection as it was in the case of Mr. Johnson.

A keeper also performs a valuable function in science. In some aspects, this individual might appear to be the least agreeable of the psychological types, because this is the person who points out your misuse of a term and seems to enjoy spotting others' errors. I once had a colleague who read one of the more prestigious physics journals, in which the maximum number of pages a paper could have was four. He would read this journal each week with an eye tuned to finding mistakes. When he did find a mistake, he would write a letter to the editor pointing out the error and indicating how it should be corrected. The errors he sought were the misapplications of theory, logical inconsistencies, and subtle mistakes in mathematics. The mistakes were not trivial and the corrections were difficult to make, so my colleague was performing a useful service despite its difficulties. Yet, he was neither tactful nor apologetic in what he did. As such, although he was smart and did good physics, he had few friends and was not well liked. Unfortunately, at some point, he became trapped in this one mode of doing science.

Creepers are brilliant people. They do not have the magic of the leapers, but they build the infrastructure that enables the rest of us to understand science. I mentioned Mark Kac earlier and his remark about some scientists being magicians and others being only extremely clever. Mark retired from Rockefeller University in the mid 80s and took up a position at the University of Southern California. He would fly to San Diego and visit the La Jolla Institute weekly, where I was Associate Director. Since I lived nearest to the airport, I would pick him up and take him to La Jolla, which I did with pleasure. Mark was a short, round man, with a white fringe around his shining cranium. He had a perpetual smile on his face and was always on the verge of laughing, particularly about some problem in science. He remarked more than once that his entire career had been focused on understanding what we meant by statistical independence. One of his remarkable papers, frequently quoted after half a century, is "Can you hear the shape of a drum?" He was a physicist/mathematician who pursued new problems the way some people go after a good meal, with appetite and good fellowship. The connections he made in statistical physics and quantum mechanics are continually being rediscovered by new generations of physicists. He avoided the fashionable and modern, preferring to stay with the fundamental. He thought about such things as the central limit theorem, how it is violated and what might be the implications of such violations. His eyes would dance when he talked about the distribution of eigenvalues in quantum mechanical systems, but no more than when he talked about music or art.

One of the connections made by Kac was between the traditional law of errors and a new law of errors that concern phenomena for which the mean value is not the best representation of the experiments. We shall examine the implications of some of his ideas in this chapter, when we look at various ways the central limit theorem can break down.

I will refrain from relating personal experiences I have had with leapers, because I have collaborated with many scientists and each would want to know why they have not been used as an exemplar in this category. For a related reason, I will also avoid the reaper category and the rest with my comments regarding Einstein.

This chapter attempts to break away from the traditional average health indicators used in the past few hundred years and anticipates how they could be replaced with more sensitive measures. The replacement measures are based on fluctuations in physiological variables rather than on their average values. The assumption was made 200 years ago that fluctuations are errors and therefore contain no information about the underlying system; this is the consequence of scientists functioning as keepers. Part of the reluctance of physicians to be leapers, or even creepers, may be traced back to their professional oath: "First do no harm." This dictum often directly contradicts what scientific research indicates needs to be done; not to do harm, but to take a chance that might ultimately result in harm. This is a discussion that would most profitably be made by those looking for the ethical implications of one kind of decision over another; weighing the potential for harm against the possible long-term good. I lift the discussion out of this ethical labyrinth by simply ignoring the issue, but still acknowledging that it is there.

Most working scientists including myself are uncomfortable discussing questions of ethics and morality as they relate to the scientific enterprise. This is due largely to the fact that the discussion is often not about the validity of the position taken in the argument, but rather it centers on the verbal ability of the person arguing, i.e. what the Greeks termed rhetoric. A practiced litigator, for example, can make the informed guess of a physician who appear to be the reckless gamble of an uncaring Doctor Frankenstein. Part of the fallacy in such discussions is the confusion of the ideal with the practical. The reality is the uncertainty of the situation in which the physician makes a decision with incomplete information. The argument is then made against the decision, based on the failure to predict the outcome, using the ideal of never making any decision in the absence of complete information. The inability of a scientist to articulate the way in which decisions are made, under conditions of uncertainty or incomplete knowledge, can always be made to appear to be either due to incompetence or negligence, particularly in the case of life and death decisions.

I now leave the often confusing domain of ethics in science and proceed to the clarity of data and its interpretation. Eventually, I examine the influence of fluctuations on physiological variables and attempt to understand why it has taken so many generations for scientists to recognize the importance of fluctuations in the determination of a patient's state of health. However, in order to accomplish this, there is a certain amount of background that must be filled in before I get to these medical data. Therefore, I examine data from a number of different complex phenomena. However, before I consider the data, let me present the way in which some ethical discussions do take place in the quiet of a colleague's office.

4.2. The Emperor in exile

In an earlier chapter, I mentioned meeting Dr. Jonas Salk in the early 80s and becoming good friends. Since his name and achievements are so well known, I thought it would be interesting to descibe the man and explain briefly his scientific status when I met him. This is also an example of being both successful and unsuccessful in science at the same time. As he was told by the legendary journalist Edward R. Murrow:

> Young man, a great tragedy has befallen you. You have lost your anonymity.

During the decade of the 1980s, I would often visit the third floor of the Salk Institute, a building constructed from the donations of the March of Dimes, where Jonas had his offices. I now derive a certain pleasure in knowing that the few pennies I donated as a child were part of the torrent of funds that manifested as two parallel rows of laboratory buildings separated by a marble court yard, that was the *Salk Institute for Biological Sciences* in the early 80s. A narrow straight line of running water, carved into the marble, ran from the wrought iron entrance down the full length of the Institute, to the concrete waterfalls at the far end of the courtyard. The line of water formed a spear that balanced on its tip the sun setting into the ocean every evening.

The laboratory buildings were constructed from poured concrete with no attempt to disguise their manufacture. Each floor was an open bay, with all the connections necessary for a laboratory channeled through the ceilings. The internal walls of the individual laboratories were movable so that their expansion and contraction could be accomodated with minimal reconstruction. The offices of Dr. Salk, himself, were fitted into a large bay on the third floor at the back of Institute, with a view of the ocean and the glider port on the cliffs high above the water on the adjoining property. The huge open bay was partitioned by textured wooden panels and doors breaking up the opening into a spacious hall running the length of his personal office, the office of his secretary of 30 years, an office for the occasional transient visitor, and ending into a large conference room at the opposite end of the bay. The walls throughout were either poured concrete or ceiling to floor wood panels, with only paintings to offset the austere setting.

Jonas' office formed a setting for a number of paintings by his wife, the internationally known artist, Françoise Gilot. One evening at a dinner party in their home, Françoise confided to my wife Sharon that it was much better for an artist to be married to a scientist than to another artist. This is because Sharon is also an artist and they were comparing notes on the compatibility of artists and scientists. They agreed on the general premise and thought it interesting that most people found the compatiblity between art and science, or at least between artists and scientists, to be peculiar. The discussions on these evenings invariably turned to the many forms that scientists and artists use to know the world, often concluding with the observation that the methods of art and science are not so different. One conclusion that stays with me is that artists and their art often anticipate scientists and their science by decades; in fact, the former may often enable the latter.

There were three windows, in Jonas' office, each nearly three feet wide and probably six feet high, starting from slightly above an observer's waist to the ceiling. The back-wall windows overlooked the Pacific Ocean, beyond a 100-foot cliff that sloped down and out, away from the Institute. There were two couches with coffee tables and accompanying chairs, along with one desk and a number of tables supporting various awards Jonas had received over the years, as well as a single book case. The room was generally Spartan in appearance, with only the paintings on the walls, the book case with volumes from all of the sciences and his white lab coat hanging on a clothes tree to indicate the taste of the man.

We would sit together in his office and have the kind of conversation that almost everyone can recall from their youth, whose memory is particularly acute since I was no longer young at that time. During undergraduate days, there would be endless discussions be it in my apartment or someone else's. Profound questions such as sin and the nature of being human would usually creep into the conversation at about 2 a.m. Of course a similar experience is also shared by those who argued the existence of love, war, justice and patriotism in a corner bar on Friday evenings. Remember, these earlier events occurred in the 60s. I recall with some sadness how infrequent such discussions became as I got older, not because I had found the answers, but because I found that I was rethinking the same old arguments without making much progress. Together, Jonas and I were able to break some of those old cycles by applying the same kind of thinking that is so successful in science and focus on problems that one actually has the hope of solving. This book is inspired partly by those discussions.

The first afternoon I visited Jonas, he suggested that we recorded our meeting as an experiment in conversation. He wanted to see if the dynamic interaction between two scientists of different backgrounds, but who are interested in the same kind of questions, would produce something new. The following excerpt is taken from a transcript of the middle of that meeting:

> *Salk*: Right. So there are lots of data (you know) to verify our suspicions, and if it can be systematized and formalized in some way, then we begin to see what kind of dynamics need to be introduced into life in order to make for a richer life. The opposite is the equivalent of, let's say, paralysis. What I think is happening in many lives is that there's the equivalent of paralytic polio, the state of the nerves being destroyed, failing to be appropriately activated, and therefore, you have a flat rather than a rich personality.
>
> *West*: Yes. I think that there exist social problems that are associated with the fact that people have elevated comfort to a goal that has preeminence. You know, comfort is more important than most other things in their lives and what most people don't realize is the logical extension of their position. If you are suspended in a fluid with all your nerves disconnected and fed intravenously, that would be the optimally comfortable situation. Yet, it simulates nothing more than death. That comfort is not a thing to be desired, particularly not for its own sake. What is to be valued is tasking yourself and the accomplishment of those tasks, and that's ... and again that is part and parcel of the evolutionary theme. You set a task, you achieve it, you accomplish it, you set a new task, you

accomplish it, and you evolve, you strengthen, you grow. I think that is what life is about.

Salk: You're constantly challenged.

West: That's right.

Salk: And if that were not true, then this complex organism with all this complexity of hearing and seeing, the various organs, various molecular-cellular systems, would never have emerged.

West: Yes, that's right.

Salk: So that the emergence of this complexity, including now the complexity of the mind that we are testing and challenging, and confronting adversity instead of retreating from adversity, instead of using drugs to suppress the feelings of anxiety we have, and finding out what that discomfort is due to, and make a necessary adjustment to find what is the evolutionary path which we may have gotten off, which is responsible for the signals of danger. So, when the discomfort level is high, then one wants to reduce it, but one wants to reduce that level of discomfort in a way that is evolutionarily valuable to us, not simply to reduce the discomfort for the sake of reducing the discomfort. *In fact, the discomfort is a signal to which one needs to respond to!!*

A little bit later in the conversation....

Salk: Now you can see if one were to approach this from the point of view of theoretical medicine, how much could be deduced and how much could be learned about metabiological medicine, as well as biological medicine. Some of these malfunctions may be biologically treatable, and that's fine. I mean, just as we immunize, we change the population distribution in terms of their reactivity to an organism, so that the dynamics of the interaction between immune system and organism, shall we say, is shifted because they have different population distributions. Now let us assume for the sake of argument, you will be able to have a population group go through an immunizing experience or immunizing education, they would then behave differently from the population untreated, unprevented, shall we say, not treated prophylactically.

Therefore, if this is true, then it is self-evident that one ought to be able to develop medicines of a biologic or metabiologic nature to treat or to prevent disease, dis-ease, illness, mal-functioning, premature death, increasing at the same time effectiveness, productivity, and all of the other things that represent the fulfillment of the biological and evolutionary potential that is undoubtedly genetically programmed to a much greater degree than has as yet been manifested. Therefore, the idea of theoretical medicine is of tremendous appeal to me, and I think could have enormous appeal in the world because it would increase the armamentaria from what we normally think of as medicine, which is biological medicine, to what I call metabiological medicine, which is the alternate holistic devices and methods and schemes which have varying scientific bases. It would be possible then, you see, to develop a scientific holistic medicine, if you like, which is the same as that of a scientific approach to metabiological medicine, which is another way of maintaining health.

West: It seems to me that the way healthcare programs work now are primarily logistic programs intended to satisfy medical needs as they should arise... that a healthcare program should in fact be a preventive program involving the application of theoretical medical concepts. In addition, this (the Salk Institute) should be the place where theoretical medicine is conducted....

Salk: Yes. Health enhancement and its health-related activities rather than maintenance, that's enhancing the positive rather than reducing the negative.

West: That's right. It's not a maintenance program.

Salk: Maintenance in the sense of repair.

West: Yes, right. You wait till it goes wrong and then you fix it.

Salk: That's right. So it's an entirely different philosophy, different orientation. My contention is that we are now emerging out of the period of time when our preoccupation was with disease and premature death, and now we should be in a phase which is much more appropriate from an evolutionary point of view, in terms of our knowledge, our technology and our comprehension, into a period that we might describe as *prohealth* rather than *antidisease*. Now that doesn't mean that we exclude antidisease, it just means that we add prohealth; so we're not taking anything away, but we're adding something to deal with what's needed, because there is now less need for dealing with problems of disease, even though we continue to deal with problems of disease with this enormous, gigantic, gargatuan, dinosaurian approach — it's out of proportion in terms of scale, ethics, morality, human dignity, and human values. All this has to do, it seems to me, to be the question, the challenge, in the light of what else we can do and how else we can approach this problem taking an evolutionary approach to it by saying that at this point in time, the situation is different from some previous point in time. It's different in Africa, in India from what it is in Scandinavia and the United States. Now when one begins to look at these problems in this way, then we begin to get some insight into what needs to be done.…

That is how the conversation went in January 1984.

Jonas Salk was small in physical stature, perhaps five feet seven inches tall, weighing approximately 140 pounds. A lean man with snow-white hair that was sparse at the crown, but which was swept back in wavy traces along the sides. In an earlier age, he would have been considered dapper, with tailored sport coats, monogrammed shirts and ascot. His face was lean with penetrating eyes that never wavered, but lightly danced over your face as he watched for your reaction. He always wanted to know what you thought and how you felt. His door was always open to visitors and his greeting came with a smile, a warm handshake, or if we had not seen each other for a few days, a hug. His movements were thoughtful and deliberate, as if to avoid startling his guests.

Jonas loved words. In the middle of a conversation, he would begin playing with a word or a phrase that had caught his fancy or that he recalled had a special meaning for him or someone he knew. What made his ruminations different from rambling was that he would try and make new connections. Occasionally, he seemed to be saying something he had said in a conversation of a week or so before, but it would be a little different; a revisiting of an old exchange, rather than a repetition of it. When I first met him, I attributed this particular habit to age (he was probably the same age then as I am now). Eventually, I learned that he had developed certain habits that helped him to keep the world in focus. One of these habits had to do with how one responds to what the world is not ready to accept. He mentioned that if the world rejects your ideas, such rejection may have nothing to do with the quality of the ideas. The failure to accept may be a consequence of the world not being ready for the ideas. Consequently, one should not give up, but rather one should continue to present the ideas to the world until the world is ready to accept them.

For Jonas, the Salk Institute was one of those ideas for which the world was not yet ready. When he started the institute, he wanted it to be a "sanctuary for odd ducks;" a scientific refuge where the novel and the iconoclastic would be common; a place where a scientific maverick could feel at home and form a community. He confided to a few people that the pressure from the Board of Directors to make the Salk Institute a traditional laboratory, had built up over the years, and eventually became so great that he was forced out as Director. In the early 80s when I met him, he was essentially relegated to his offices with no laboratory and little or no influence on the daily workings of the institute that proudly bears his name. He had the honorific title as the Founding Director. His reaction to all this was not animosity towards the Board, nor harsh words for the then Director of the Institute, who was a physicist. He had concluded that the world was not yet ready for a research institution that does not do science in the conventional way.

One of the areas of research on which we collaborated was the scientific study of creativity. The view we shared was that creativity was a complex process that neither my background in the physical sciences, nor his background in the biological and medical sciences, nor our contact with the arts through our wives, had adequately prepared us to study. We had a few nascent ideas, arising from the overlap of our various disciplines, but we thought that to attain anything significant, it would require a collaborative institution which is not unlike his original vision of the Salk Institute. We both knew that obtaining funding for such an activity would only be successful if we could articulate our vision of creativity and what we hoped to accomplish through such an institution. Putting the vision into words was particularly difficult because there was so much nonsense disguised as intellectual achievement in the academic world and most value-laden words had been turned from gold into lead. We were never successful in raising the money necessary to extend the Salk Institute into this area of study.

4.3. What is wrong with the law of errors

The law of errors, like most successful theories, has a number of fathers. The mathematics-father of the bell-shaped curve was de Moivre, who constructed it from a simple two-outcome game of chance, such as flipping a coin. The physics-father of the law was Gauss, who constructed it to understand the nature of experimental variability in astronomical observations. Neither father anticipated the profound effect that this somewhat arcane mathematical construction would have on civilization. To understand this influence on society, let us peer beneath the mathematical veil of this off-spring of science and explore the implications of the law of errors. Eventually, what we want to understand is whether or not medical phenomena satisfy the criteria for the application of the bell-shaped curve, but before we can make that determination, we are obligated to examine the properties that a system must have in order for the law of errors to be applicable to its description.

I examine specific phenomena to understand when the law of errors works well and when it breaks down. Consider why is it that

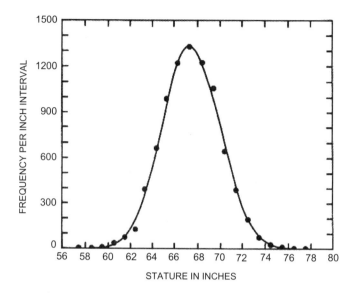

Fig. 4.1: The points indicate measurements of the number of adult males in the British Isles having the indicated height in inches. The solid curve is the Gauss distribution fit to the data. (Taken from Yule and Kendal.[33])

a genetically determined process like the distribution of heights satisfies the bell-shaped curve, as shown in Fig. 4.1. The height of an individual is determined theoretically by adding together a number of bones of different lengths. The foot is connected at the ankle to the shin, which is connected at the knee to the thigh, which is connected to the hip and so forth. The random variations in height, variations produced by small changes in the length of each bone in each individual in a human population, is given by the law of errors. The variation in the lengths of the bones is produced by the genetics of the individual and is a phenotypic characteristic of the person. The bell-shaped curve for a phenotypic characteristic, such as the distribution of height, can be ascribed to a genetic factor only if this factor operates *randomly* and *independently* on each of a large number of genes, which work together to produce the phenotypic characteristic of interest. In this case, the physical size of an individual is determined by the sum of the sizes of more than 200 bones.

In a large population of adult males, such as in the British Isles example used in Fig. 4.1, the small accidental differences in the average size of a particular bone fluctuate randomly from person to person, and independently from bone to bone over the population. These variations are produced by a large number of both environmental and genetic factors. Some bones are longer than the average length for that bone in the population and some are shorter, so the shorter and longer tend to nullify one another in determining the average over the population. However, the degree of variability, i.e. how much longer and how much shorter, is determined by the width of the distribution, or as we said earlier, by the standard deviation.

Consider a somewhat simpler example in order to better appreciate how one or two numbers, such as the average value and standard deviation, can lead to the understanding of an apparently complex process. Take the example of a large class in Existential Philosophy and poll the students to assess the professor's teaching ability by grading her on a scale of 1 to 10. The results come back and the average ranking she receives is 8.5. On an absolute scale, this is a good grade, a solid B. Now construct two distinct scenarios. In the first case, the highest grade obtained is 9.5 and the lowest is 7.5. From this, I might conclude that the students essentially liked the professor's teaching, since there is relatively little variation in the grade. In the second case, suppose the student's rankings span the full range from 1 to 10. Note that there would be many higher grades in order for the average to reach 8.5, but there must also be a significant minority that strongly objects to the way the course is being taught. Without making a value judgment about which teaching style is preferable, the professor with greater variability in scoring is certainly more controversial, since the students' responses are so varied. Thus, these two situations in which the average student population thought that the professor is a good teacher are really quite different. The single number, the average, is not adequate to give a clear picture of the professor's teaching. A second number, the degree of variability in the students' responses, and the width of the distribution are able to give us more insight. The same is true for the distribution of heights.

Our discussion of the mean and the standard deviation (width of the distribution) has to do with the fact that the outcome of any particular experiment has little or no significance in and of itself. Only the collection of measurements, the ensemble, has significance and this significance is manifested through the distribution function, the curve in Fig. 4.1. The distribution function, also known as the probability density, associates a probability (relative frequency) with the occurrence of an event in the neighborhood of a given magnitude event. What this means for the present data set is that if I closed my eyes and stuck a pin in a large map of the British Isles, went to that location and selected the first person I met, the bell-shaped curve will give the probability that the randomly selected individual will have a given height. Yet, this interpretation only makes sense if I do my map sticking numerous times, because the distribution does not apply to a single pin prick. From this distribution, it is clear that the likelihood of encountering a male six feet in height on my trip to Britain is substantial, that of meeting someone six feet six inches tall is much less so, and the probability of seeing someone greater than ten feet tall is virtually zero.

So what is it that I can extract from the discussion on the distribution of heights that concerns all phenomena to which the bell-shaped curve applies? The first property is that the height of any individual in the population does not influence the height of any other individual in the population. This is of course not true in families where genetic traits are passed from generation to generation, but the interdependence of a few individuals in a population of millions is negligible, so each fluctuation in the height is considered to be statistically independent of any other fluctuations in the height. In a more general context, each measurement error is independent of any other measurement error; each experiment stands on its own and the outcome of a given experiment does not influence the outcome of any other experiment.

The second property in the measurement of height is that the individual fluctuations add to one another. If there is an error of a given size in the shin, it adds to any error in the femur, which adds to any error in the backbone, and so forth. We can think of each bone in the body of the average man (woman) as having its own average value, resulting in the peak of the bell-shaped curve for the height of the average man (woman) in the population. However, in a real person, there are fluctuations around each average bone length, some positive and some negative, with the sum of fluctuations producing variations in the overall height. Consequently, errors are additive.

I see no reason not to assume that the biological mechanism determining the length of each bone in the body is the same, since all members of a given population have shared characteristics. Given this assumption, it is also reasonable to presume that the fluctuations in the length of the bones are all of one kind. So the physiological mechanism that produces a fluctuation in the length of the shinbone is the same as that which produces a fluctuation in the femur. Note that we are not discussing pathological fluctuations but only the normal variability observed in a healthy population of adults. This property is summarized in the statement that the variations all have the same kind of statistics, which is not unreasonable as long as I compare like to like.

The final property necessary for the application of the central limit theorem to our errors is that the width of the statistical distribution of the individual errors is finite. This requirement implies that the width of the underlying distribution, whatever it is, cannot increase without limit. In the case of the distribution of heights, it would be unreasonable to have a distribution whose width is greater than the average of the distribution (peak of the bell-shaped curve). In Fig. 4.1, the peak of the curve is located at 68 inches and the width of the curve is approximately 8 inches, which is actually twice the standard deviation. Therefore, a person 6 feet 4 inches would be considered tall, but not unreasonably so. A person of 7 feet would be considered extraordinary, e.g. an NBA basketball player, but he would not be grotesque. Note that I have only gone out to an extreme of height that is twice the width of the distribution. If I go much beyond this extreme value, I would be entering the world of the pathological.

Thus, the world of the bell-shaped curve is one in which fluctuations are many; they are statistically indistinguishable from one another; they add together and they never get too large. Note that a fluctuation is defined by the amount that the measurement deviates from the expected or average value in a large number of measurements. This is the world of the central limit theorem and it is considered by many to be the one in which we live in. Many, if not all, the variations measured in physiological phenomena are considered to be subjected to the law of error, just as the measurements were in the physical sciences.

In the world of the central limit theorem, the average heart rate is the best way to characterize the cardiovascular system. The variability in the instantaneous heart rate would be due to random fluctuations that are associated with the complexity of this system, but these fluctuations would be assumed to carry no information. In such a world, these fluctuations would satisfy a bell-shaped curve.

In the world of the central limit theorem, the average breath rate indicates the state of health of a person's respiratory system. The variability in the instantaneous breath rate would be due to unpredictable and unimportant rapid changes in the outputs of the various subsystems of which the respiratory system is made. These physiological variations in the breath rate would not be associated with any control mechanisms and therefore they could be ignored with impunity. Again, in this world, these fluctuations would also satisfy the law of errors.

In the real world, it is not possible for me to meet someone who is twice my height, assuming that I fall within the bell-shaped distribution of heights. Yet, is this distribution really indicative of the complex phenomena in the world around us? Is the distribution of income in any country given by a bell-shaped, curve? The answer is a resounding "no". The truth of this observation can be determined by using only personal experience without surfing the Internet and downloading any data.

It is reasonable to conclude that I will not meet someone who is twice my height, because of the bell-shape of the height distribution. On the other hand, it is also reasonable that I could meet someone who has twice my income, e.g. my boss when I worked for a small company, or 10 times my income when I met other scientists who worked for large industrial laboratories. However, I cannot rule out the possibility of meeting someone with 100,000 or even 1,000,000 times my income, e.g. Bill Gates at the airport. It is therefore clear that the distribution of income is very different in structure from the distribution of height. Unlike height, where the measurements cluster in the vicinity of the average value, income broadly spreads out with some individuals at the most extreme levels. The region of the distribution where the large values of income exist is known as the tail, because the income levels are so far from the central region, where the mean income of ordinary wage earners resides. Furthermore, because these large levels of income occur in such a significant fraction of the population, they are sometimes known as heavy-tailed distributions, or for more technical reasons, they are termed inverse power-law distributions.

4.4. The inverse power-law distribution

The fact that the distribution of income has an inverse power-law form was first established by the Italian engineer/economist/sociologist, Vilfredo Pareto, towards the end of the 19th century. In Fig. 4.2, the income distribution for people in the United States in 1918 is depicted on graph paper where both axes are in terms of the logarithm of the variable. On the vertical axis is the logarithm of the number of individuals with a given level of income and on the horizontal axis is the logarithm of the income level. The closed circles are the data showing the number of people with a given level of income and the solid line with a negative slope indicates the Pareto law of

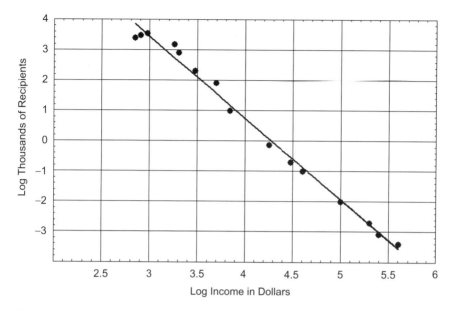

Fig. 4.2: Pareto's Law: The frequency distribution of income in the United States in 1918 is plotted on log-log graph paper. A down going straight line on this kind of graph paper indicates an inverse power law and the slope of the line gives the power-law index.

income.[30] It is from such curves that Pareto sought to understand how wealth was distributed in western societies, and from this understanding, to determine how such societies operate. Unlike most of their 19th century contemporaries, Galton and Pareto believed that the answers to questions about the nature of society could be found in data and that social theory should be developed to explain information extracted from such observations, rather than being founded on abstract moral principles. Again, the conflict between the ideal and the practical is evident.

If I divide the number of individuals in a given income interval, the number on the vertical axis in Fig. 4.2, by the total population of income earners of the country, the curve denotes the fraction of the population having a specific income. This is a relative frequency curve which can be interpreted as a probability density. As a probability curve, Fig. 4.2 can be directly compared with the Gauss distribution, as we did in Fig. 3.6. Notice that the Pareto curve stops abruptly at

$600, i.e. the "wolf-point," the point at which the wolf is at the door. In 1918, the income of $600 per year was the minimum level of income necessary to maintain life with a modicum of dignity.[2] Pareto found that many western societies had inverse power-law income distributions independently of their particular social organizing principles. He was the first of many scientists to find the inverse power law rather than the bell-shape curve, in the analysis of real-world data from many different kinds of complex social phenomena.

In itself, the construction of a distribution of the inverse power-law type would not have been particularly important. Part of what makes the Pareto distribution so significant are the sociological implications that Pareto and subsequent generations of scientists were able to draw from it. For example, he identified a phenomenon that was subsequently known as the *Pareto Principle*, i.e. that 20% of the people owned 80% of the wealth in western countries. It actually turns out that fewer than 20% own more than 80% of the wealth. The actual numerical value of the partitioning is not important for the present discussion. In any event, the 80/20 rule has since been determined to have application in all manner of social phenomena, in which the few (20%) are vital and the many (80%) are replaceable. The phrase "vital few and trivial many" was coined by Joseph M. Juran in the late 1940s, and he is the person who invented the name *Pareto Principle* and attributed the mechanism to Pareto. This rule has recently caught the attention of project managers and other corporate administrators who now recognize that 20% of the people involved in a project produce 80% of all the results; that 80% of all the company problems come from the same 20% of the people; resolving 20% of the issues can solve 80% of the problems; 20% of one's results require 80% of one's effort, and so forth.

Consequently, with the 80/20 rule in mind, complex systems (phenomena) suggest that managers should concentrate on the 20% that truly makes a difference and only lightly monitor the 80% that do not significantly influence the operation of the system. I will elaborate on this point in subsequent discussions of physiologic systems, but for the moment, let us satisfy ourselves with examples of natural and social phenomena that are described by inverse power-law distributions, and as a result, the underlying phenomena are subjected to the *Pareto Principle*.

I argue that the inverse power-law distribution is ubiquitous in the real world of complex phenomena and not the bell-shaped curve; furthermore, the central limit theorem must be modified to accommodate this reality. However, let me first review a few more examples of complex social and biological phenomena that are described by such inverse power-law distributions, so that the reader has some perspective on just how pervasive this description of complexity really is.

Consider the phenomenon of speciation in biological evolution, that being the biological variability associated with a given genera. Species are commonly taken as the fundamental elements in the classification of living organisms. The species are then combined into coarser classification groups called genera. The distribution in the number of species in a genera is given by an inverse power law of the Pareto form, when the data are ordered in rank. This distribution was found from the data by Willis[34] at the beginning of the last century. Willis collected data on various families of plants, animals and insects, and graphed the logarithm of the number of genera as ordinate against the logarithm of the number of species in each genus as abscissa. By analogy with Pareto, the number of genera is associated with the number of people and the number of species is associated with the level of income. Again, in biology as in economics, on doubly logarithmic graph paper, the data graphed as a straight line indicates an algebraic relation between the two variables, an inverse power law. The distribution was given in rank-ordered form, with the most frequent first, the next most frquent second, and so on. The monotypic genera, with one species each are the most numerous, the ditypics, with two species each, are next in rank, and genera with higher numbers of species becoming successively fewer in a rank-ordered way, (see Fig. 4.3). The inverse power-law relation depicted in this figure was explained by Willis to be due to random biological mutations[34]:

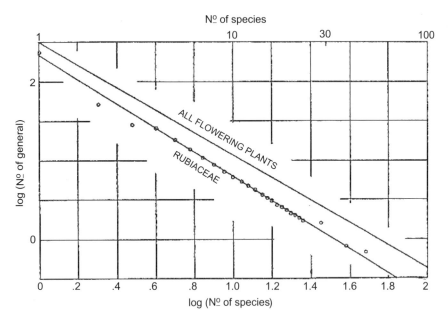

Fig. 4.3: Willis' Law: The solid line compares the number and size of *Rubiaceae* with the curve for all flowering plants. (Taken from Lotka.[35])

If species of very limited area and genera of one species (which also have usually small areas) are, with comparatively few exceptions, the young beginners in the race of life, and are descended in general from the species of wider dispersal and the larger genera, and if the number of species in a genus is, broadly speaking, a measure of its age, the idea at once suggests itself that a given stock may be regarded as "throwing" generic variations much as it throws offspring, so that the number of genera descended from one prime ancestor may be expected to increase in geometric ratio or according to the law of compound interest. The number of species descended from one ancestor might be expected to follow the same form of law with a more rapid rate of growth. On such a very rough conception it is found that the form of frequency distribution for sizes of genera should follow the rule that the logarithm of the number of genera plotted to the logarithm of the number of species gives a straight line.

These properties are today explained at the level of molecular mutations in the genes, where we also find fractal properties across time in the number of molecular substitutions within a protein.[36] Note that the inverse power law in Fig. 4.3 was found to be true for all flowering plants, certain families of beetles, as well as for other groupings of plants and animals. The inverse power law is representative of literally hundreds of such plots that have been made on botanical and biological data since the initial work of Willis over a century ago.

It is not only the complexity found in economics or in biological evolution that leads to inverse power laws. The distribution in the number of cities having a given population was determined by Auerbach[37] to be of the same form as the law of Pareto. In Fig. 4.4, the "cities" in the United States from 1790 to 1930, with populations of 2,500 or more, are ordered from largest to smallest; a city's rank is its ordinal position in the sequence. In this way, Auerbach obtained a straight line with negative slope on doubly logarithmic graph paper. In Fig. 4.4, the inverse power-law behavior of the distribution of cities does not change in time, but rather the slope of the separate curves do not change in time. It is apparent that the distribution of communities in 1790, the lower left part of the figure, is parallel to that same distribution in 1930, and more or less with all the curves in between.

Thus, it would appear that the underlying mechanism of urban development, regardless of what it is, as measured by the power-law index, has not changed throughout the indicated 150-year history of the United States. One possible mechanism for the growth of a city regards its desirability, the manner in which a population center attracts youths from the farms and from other countries. A boy from the rural area would be attracted to the largest city nearby, often mistaking size for the opportunity to make one's fortune. If not riches, one could find work and make a living in a city. This may have been the attraction that cities held, not just for country youths, but for European immigrants as well.

It seems that we have introduced two distinct kinds of inverse power-law distributions here. One is determined by a rank-ordering

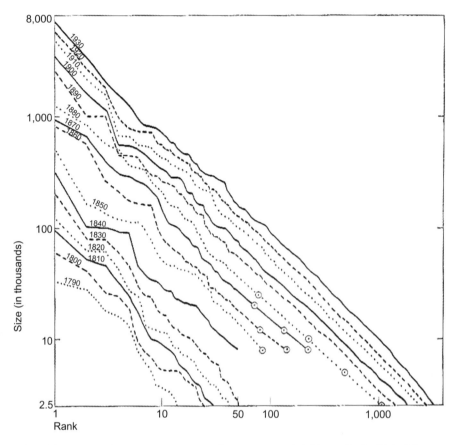

Fig. 4.4: Auerbach's Law: Communities of 2,500 or more inhabitants in the United States from 1790 to 1930, ranked in decreasing order of population size. (Taken from Zipf.[39])

procedure, Willis' Law of speciation and Auerbach's Law of urban growth; and the second is determined by the frequency distribution, Pareto's Law of income. Adamic[38] explains that these are in fact not two separate kinds of inverse power laws, as can be seen from a more careful articulation of the rank-ordered distribution. The equivalence is most readily seen by noting that if R is the rank of the city with N inhabitants, it can also be said that R cities have N or more inhabitants. The latter phrase expresses the same form of relation as that established by Pareto when he relates R individuals to the level of income N. The ordinate and abscissa have been interchanged for the two cases, for Pareto the independent variable is N (x-axis) and the dependent variable is R (y-axis), whereas for Willis and Auerbach, the reverse is true. Given this interchange of dependent and independent variables, it is readily determined that the power-law indices in the two cases are the reciprocal of one another. Thus, there are not two distinct kinds of inverse power-law distributions, but rather there are two ways of expressing essentially the same relation between variables in complex phenomena.

The final example given here is perhaps the most well known of the inverse power laws and has to do with how frequently words are used in a language. If the reader is interested in more examples, he may consult *Physiology, Promiscuity and Prophecy at the Millennium: A Tale of Tails*,[32] in which the "tails" in the title are the tails of the inverse power laws in the data. In Fig. 4.5, the most often-used

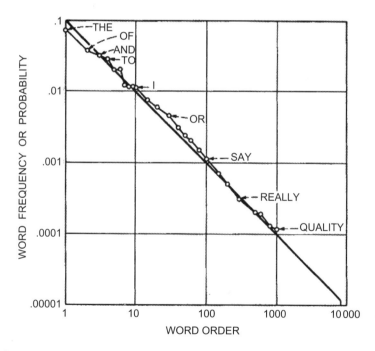

Fig. 4.5: Zipf's Law: Word frequency is graphed as a function of rank order of frequency usage for the first 8,728 words in the English language, using the text of Ulysses to obtain data.[39]

word in James Joyce's novel, *Ulysses*, is graphed first, the second most frequently used word in that novel is graphed second and so on. The 260,430 words in Ulysses were counted by Dr. M. Joos, as were the number of distinct words, 29,899, and their order of ranking in the sequence of words. George Kingsley Zipf (1902–1950), a Harvard University linguistics professor, took these data and showed they gave rise to an inverse power law, which has come to be known as Zipf's Law. In fact, it was through Zipf's seminal book *Human Behavior and the Principle of Least Effort*,[39] published at his own expense, that the ubiquity of inverse power laws becomes evident. Zipf reviewed all the inverse power laws known at that time and extended their interpretation to natural language. He showed that the inverse power law is a property of all natural languages, the difference between languages being manifested in the magnitude of the slope of the distribution graphed on doubly logarithmic graph paper. However, the full implications of Zipf's work are still being developed and the inverse power-law distribution is often referred to as Zipf's Law or Pareto's Law, depending on the background of the investigator. In fact, these laws refer to the same deep relations within the systems being studied.

Today, it is difficult to imagine the dismissive resistance presented by the social scientists to Zipf's observations. In this regard Mandelbrot offered the following:

> ...The most diverse attempts continue to be made, to discredit in advance all evidence based on the use of doubly logarithmic graphs. But I think this method would have remained uncontroversial, were it not for the nature of the conclusion to which it leads. Unfortunately, a straight doubly logarithmic graph indicates a distribution that flies in the face of the Gaussian dogma, which long ruled uncontested. The failure of applied statisticians and social scientists to heed Zipf helps account for the striking backwardness of their fields.

The distribution in the frequency of the use of words was taken up by Shannon in his application of information theory to the English language.[10] What could be better in discussing information than how to employ the complexity of language in communication? Recall that one bit of information is gained each time I toss a coin and observe the outcome. If I assume that the 26 letters of the alphabet and a blank were equally probable, such as the sides of a multifaced cube, only that there are more than six faces, I would be able to determine that the information gained is 4.76 bits/symbol. However, I know that certain letters are used more often than others, for example, *e* certainly occurs more frequently in English than does *q*. Consequently, if I use the actual relative frequencies of the individual letters in English, the information gained in a long sequence of letters decreases from 4.76 to 4.03 bits/symbol.[40] Individual words, like individual letters, do not appear as randomly as a sequence of coin tosses. In order to convey meaning, word sequences are correlated and the frequency with which words are used are constrained. The inverse power-law distribution for the frequency of word usage implicitly contains those constraints. Using Zipf's law, it is possible to determine that in English, the information gained from the pages of a book is 11.82 bits/word. On the other hand, by adding the constraint of 5.5 letters per word on average, it results in a decrease in information to 2.14 bits/letter, almost 50% of what it was above. Information theory, therefore, provides a formal way to compare the level of complexity of various modes of communication.

If I can determine the information contained in a normal speech using information theory, could I similarly quantify certain mental pathologies that affect what we say? For example, is the information content of the speech of a bi-polar individual different in the manic and depressed states, and are these quantifiably different from that same person when they are normal? If so, can I get at this through simply counting the frequency of the word usage independently of the "meaning" of what the individual is saying? An example from my own experience concerns my brother, who is bi-polar, and who lived with us occasionally for 10 years. When my wife and I heard the words *Jesus*, *heaven*, or *universe* in his conversation, we knew it was time to take him to the VA Hospital to renew his medication.

The phenomena of biological evolution, writing, talking, urban growth and acquiring a fortune are statistically more similar to one another than they are to the growth of individuals in a human population. The chances of a child, picked at random from a population, getting rich and/or becoming famous as they grow older are much greater than the chances of that child becoming tall. Part of this good fortune is the difference in the intrinsic nature of the kind of processes involved.

The bell-shaped curve describes a linear additive statistical process, which is most easily remembered by means of the refrain from the old spiritual song, *Dry Bones*[41]:

> Your toe bone connected to your foot bone
> Your foot bone connected to your anklebone
> Your anklebone connected to your leg bone
> Your leg bone connected to your knee bone
> Your knee bone connected to your thighbone
> Your thighbone connected to your hipbone
> Your hipbone connected to your backbone
> Your backbone connected to your shoulder bone
> Your shoulder bone connected to your neck bone
> Your neck bone connected to your head bone
> I hear the word of the Lord.

And since there is little deviation in the average size of each of the individual bones that make up our bodies, the distribution in height is bell-shaped. In such a linear additive process, if one element is lost, the effect is relatively minor and in a large population, its loss is neither missed nor lamented.

On the other hand, the process described by the inverse power law is most easily captured by Benjamin Franklin's lyric in *Poor Richard's Almanac*:

> For want of a nail a shoe was lost.
> For want of a shoe a horse was lost.
> For want of a horse a knight was lost.

For want of a knight a battle was lost.
For want of a battle a kingdom was lost.
And all because of a nail.

This lyric suggests a multiplicative relationship of interdependent events in which everything depends on everything else. In such a nonlinear multiplicative process, if one element is lost, the entire process fails, or is at least degraded. In fact, this implicit connectedness of the multiplicative (as opposed to the additive) basis leads to long tails in the distribution function describing the inverse power-law process.

Figure 3.6 shows that the two distributions, the Gauss and the inverse power law, could not be more dissimilar. The former is confined to a relatively narrow region, being unable to capture the full range of variation observed in actual data and has been the distribution of choice for nearly 200 years. The bell-shaped curve has been used to describe everything from our intelligence to how bridges collapse. From the comparison made in Fig. 3.6, it is clear how one might think that the average value is a good representation of data, if the data satisfies the central limit theorem (bell-shaped curve). However, if the data satisfy the inverse power law, shown to be so much broader than the bell-shaped distribution, the average value is no better, and may in fact be worse than other statistical measures of the phenomenon.

Thus, the data on wealth, language, urban growth, biological evolution and apparently most other complex phenomenon exhibiting fluctuations have distributions that are more like the inverse power law than they are like the bell-shaped curve. This strongly suggests that we abandon the 200-year-old tradition of Gauss and be guided by the data in the medical sciences, as Gauss was guided by the data in the physical sciences. Therefore, the data may usually yield results greater than what the anticipated bell-shaped curve would yield as the average value.

4.5. How the physical and life sciences are different

Quetelet proposed that the average alone represents the ideals of a society in the areas of politics, aesthetics and morals. His view involved, in the words of Porter,[42] "a will to mediocrity." Like Cannon in medicine, Quetelet's notion involved a tendency for the enlightened to resist the influence of external circumstances and to seek a return to a normal and balanced state. The idea that a property of society could be defined, but not have a finite average, would have been anathema to him. After all, the notion of an average is apparently fundamental to everything we do in society. How could we get along without it?

My teacher, colleague and friend, Elliot Montroll, along with my friend and classmate in graduate school, Wade Badger, in their book *Introduction to the Quantitative Aspects of Social Phenomena*,[2] gave a number of examples of social processes that are described by inverse power-law distributions. One particularly intriguing example had to do with the number of sexual relationships that a student might have with members of the opposite sex. It was posed in the form of a question concerning the fraction of male undergraduates who have sexual relations during their collegiate life with a given number of young women. Many have relations with only one girl, somewhat fewer with two, certainly less with three, and not an insignificant number with 10 or more. As Montroll and Badger said, however, there are certain self-styled don Juans per 100 of students that play this social game, who have seduced 20, 50 or even 100 times as many girls as the average undergraduate. The latter phenomenon was documented in the *Kinsey Report* of 1948.

So how does this social interaction achieve an inverse power law in the number of males having a prescribed number of liaisons? It would be necessary for an individual to have a number of talents that would contribute to his ability to form the necessary relationships.

Montroll and Badger speculate that among these abilities might be: (1) the experience to select a girl who seems to be receptive to an approach; (2) the ability to charm the girl into accepting a first date; (3) the tenacity to maintain the relationship over a sequence of dates; (4) the perceptiveness to recognize the moment when petting[a] should be the natural next step; (5) the ability to go beyond petting and win the young woman's affection and stimulate her emotions to the point where the proposal of the crucial stage of the affair is inevitable; (6) a willingness to persist even if rebuffed, until consent is obtained; (7) the determination to find the appropriate quarters where the seduction can be completed; and (8) the ability to physically perform. The overall probability that an individual male will reach coition is, according to the model, the product of the individual probabilities of possessing each of the necessary characteristics listed. If a male's probability in each item exceeds another by 50%, his overall success rate will be larger by nearly a factor of 26 for the eight characteristics listed. In other words, for each liaison of an ordinary student, this don Juan would have 26 such liaisons. This is the extraordinary nature of the multiplicative property of the inverse power law; a relatively small change in the individual characteristics are amplified in the overall probability of success.[b]

Finally, the distribution of sexual liaisons anticipated by Montroll and Badger between members of the opposite sex have been determined as a *web of human sexual contacts* by Liljeros *et al.*[43] in a *Nature* paper of the same name. The authors of that paper find that the distribution in the number of sexual partners in a year is inverse power law for both males and females, (see Fig. 4.6). These data were obtained from a survey of sexual behavior taken in Sweden in 1996. The survey involved a random sample of 4,781 Swedes and

---

[a] Petting is a term that was popular when I was a teenager and if I remember correctly, necking involved anything above the waist and petting involved everything else, but without having sexual intercourse.
[b] It should probably be noted that these were the requisite "talents" for seduction when I was of an age for them to make a difference. These talents have no doubt changed over the years, but the form of the distribution, inverse-power law, has not.

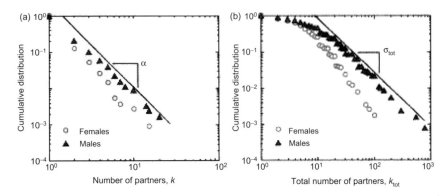

Fig. 4.6: Scale-free distribution of the number of sexual partners for females and males. (a) Distribution in the number of partners in the previous 12 months. Note that the male and female data points are roughly parallel; females have a slope $\alpha = 2.54 \pm 0.2$, while males have a slope $\alpha = 2.31 \pm 0.3$. (b) Distribution of the total number of partners over respondents' lifetimes. For females $\alpha_{tot} = 2.1 \pm 0.3$ and for males $\alpha_{tot} = 1.6 \pm 0.3$. (Taken from Ref. 43 with permission.)

had a 59% response rate. Apart from a slight skew of the sample for elderly women being slightly under sampled, the sample is representative. The web of sexual liaisons twinkles in time with different connections blinking into and out of existence. Due to the dynamic character of this web, the data are taken for the 12 months prior to the survey. It is apparent from the data that males report a larger number of sexual partners than females, but both sexes have the same form for the distribution function. An analogous distribution was obtained for same-sex liaisons.

Anderson and May[44] found that the distribution in the number of sexual partners of homosexual men in Great Britain was also an inverse power law. One consequence of this mating distribution was the existence of homosexual "don Juans," such as the Canadian airline steward Gatan Dugas, who in the book, *And The Band Played On*,[45] was reported to have gone from bathhouse to bathhouse in San Francisco, infecting young gay men with AIDS. It is by means of such amplification mechanisms that epidemics spread out of control, and which I was not ready to mention in the earlier discussion of smallpox. These kind of amplification factors are absent from the traditional studies of the errors in measurements within physical systems, not because they do not exist; but because they were not sought after, and therefore they were not found in the statistical analysis of physical phenomena until fairly recently.

From these examples of distributions with heavy tails, I draw the conclusion that the law of errors and the bell-shaped distribution are only appropriate for describing fluctuations in simple systems that are linear and additive. On the other hand, fluctuations in complex systems are more likely to be described by inverse power-law or scale-free distributions. The multiplicative nature of the interactions in complex phenomena may explain the resistance of such phenomena to certain kind of disruptions and the catastrophic failures induced by other kinds of disturbances.

One can construct a probability function from physiologic time series data in the same way as it was done for simple gambling processes, or for the distribution of the heights of males in the British Isles. In the case of heartbeat time series, the time axis can be partitioned into intervals of 100th of a second, thereby segmenting one second into 100 equal intervals. I can count the number of events corresponding to heartbeats that have a change in time interval between two successive beats, of e.g. 0.50 sec, 0.51 sec, 0.52 sec, etc. In this way, I determine the number of events (change in the time interval between heart beats) in a given time interval and by dividing this number by the total number of events, I obtain the relative frequency of events in a given time interval. The results of applying this procedure to cardiac time series are shown in Fig. 4.7.

Figure 4.7(a) shows the analysis of HRV data obtained from *Physionet* with a total of 106,999 intervals, where the variable (change in interbeat time interval) has been divided by the standard deviation, such that the horizontal axis is dimensionless. The data points appear to fall on the bell-shaped curve of Gauss, given by the solid curve. The empirical and theoretical curves do look very similar to one another. So how do we tell them apart?

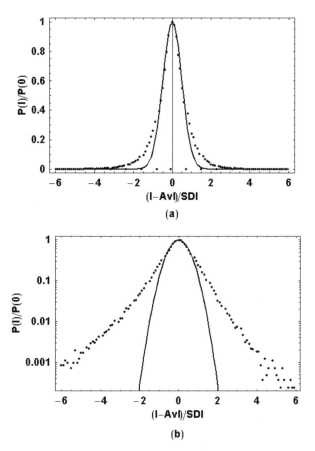

Fig. 4.7: The empirical relative frequency $P(I)$ of the differences in the beat-to-beat intervals I. To facilitate comparison, the variable is divided by the standard deviation of the increment data and the relative frequency is divided by $P(0)$. The solid curve is the best Gaussian distribution fit to the data.

If the relative frequency of the change in the interbeat interval, the vertical axis, in Fig. 4.7(a) is replaced by the logarithm of the relative frequency, it results in Fig. 4.7(b). In the latter figure, the distinction between the empirical distribution and the "best" Gaussian distribution is clearly seen. The two distributions coincide in the vicinity of the average interbeat interval, where the Gaussian distribution is supposed to be the best, but after two standard deviations, the curves are quite different. At ± 2 (corresponding to plus or minus two standard deviations in the original units) on the horizontal axis, the magnitude of the probability density between the two curves differs by a factor of 100. Consequently, if a histogram of the data predicts a 5% chance of obtaining a change in time interval of a given size, the Gaussian distribution fit to the data would predict a 0.05% chance of obtaining the same value. Thus, the two distributions deviate from one another very strongly in the tail region. Therefore, the heartbeat data indicates that the intervals for the changes in the heartbeats can be much longer than one would have predicted, based on a bell-shaped distribution.

In Fig. 4.7(b), the distribution of this physiological process and the beating of the heart leads to a distribution with a very long tail; a tail that in fact satisfies an inverse power-law distribution. The heartbeat time series is determined to be asymptotically scale-free and is therefore a fractal process, as discussed more completely in the following chapters. Furthermore, the central limit theorem, which predicts the distribution of Gauss and de Moivre, does not describe the cardiac time series. Putting this information together, it establishes the motivation for what typically fails in the application of the central limit theorem to such physiological processes. There is certainly enough data in this case, with approximately 100,000 time intervals to test any assumptions made in constructing a model. The heart beating process does not change during this time, or rather, it does not change dramatically, so the statistics of the fluctuations should be unchanged as well. The remaining properties necessary for the central limit theorem to be valid are independent of events and a finite value of the width of the statistical distribution. It is the latter condition that is set aside here.

The width of the empirical distribution is significantly wider than that of the Gaussian distribution, as evident in Fig. 4.7(b). This difference in width suggests that the statistics of the underlying processes may not have the finite width for an infinite number of data points, as would be required by the central limit theorem in order for the

Gauss distribution to represent the fluctuations. Phenomena with such strange statistics interested the equally unique mathematician, P. Lévy.

At the turn of the 20th century, there did not exist a probability calculus; instead, there was a collection of small computational problems. It was in 1919 that Paul Lévy (1886–1971) was invited to give three lectures at the Ecole Polytechnique in Paris on[46]:

> ...notions of calculus of probability and the role of Gaussian law in the theory of errors

Not given to doing anything by half-measures, these lectures began Lévy's lifelong romance with probability theory, the Gaussian distribution and the central limit theorem. His studies led to his being identified as the major influence in the establishment and growth of probability theory in the 20th century. A close friend, another leading mathematician, Michel Loève, explains Lévy's work after the latter's death in this way[46]:

> Paul Lévy was a painter of the probabilistic world. Like the very great painting geniuses, his palette was his own and his paintings transmuted forever our vision of reality. Only a few of his paintings will be described here — some of those which are imprinted indelibly on the vision of every probabilist. His three main, somewhat overlapping, periods were: the limit laws period, the great period of additive processes and of martingales painted in pathtimes colors, and the Brownian pathfinder period.

Unlike other mathematicians, Lévy did not follow the mathematical crowd. He relied on intuition to such an extent that the mathematician/physicist, John von Neumann, commented in 1954:

> I think I understand how every other mathematician operates, but Lévy is like a visitor from a strange planet. He seems to have his own private methods of arriving at the truth, which leave me ill at ease.

A view that was reinforced by another world-class probabilist, Joseph Doob, who wrote:

> He [Paul Lévy] has always traveled an independent path, partly because he found it painful to follow the ideas of others.

Paul Lévy did fit the romantic mold of the isolated scientist, working alone, thinking great thoughts, and producing scientific work that changed our global view, without his name becoming an every day household word.

In the 1920s and 1930s, when other mathematicians were looking for the most general conditions under which the central limit theorem could prove to be true, the maverick Lévy was seeking the general conditions that violate the central limit theorem, but which still led to a limit distribution. By not following the scientific crowd, Lévy was able to construct a generalized central limit theorem, and from this theorem, he constructed a new class of probability distributions. Lévy's generalization of the central limit theorem, has in the hands of his mathematical successor Beniot Mandelbrot, become one of the most valuable mathematical tools in the arsenal of the physical and life scientists for describing complex phenomena, in the latter part of the 20th century.

The kind of statistical fluctuations encountered in physiological phenomena do not have the familiar properties of the bell-shaped curve, such as having a well-defined mean or a finite standard deviation. The general distribution of Lévy always has a diverging width (standard deviation) and can even have a diverging average value. This was not such an easy concept for me to grasp when I first encountered it. It was gratifying to find that the ambiguity regarding the mean was an old problem in statistics, and in fact, it was first encountered before there was a discipline of statistics. The scientist who confronted this issue was Nicolaas Bernoulli, the brother of our old friend Daniel Bernoulli. As was the custom of the time, he presented the problem in terms of a game of chance and published it in the St. Petersburg Journal, which was subsequently named the St. Petersburg game.

Consider the game in which a player tosses a coin and a second player agrees to pay $2 if a head occurs on the first toss, $4 if the player gets a first head on the second toss, $8 if the player gets the first head on the third toss and so on, doubling the prize every time a winning head is delayed by an additional toss. What are the player's expected winnings in this game and what should be the ante? We know that the probability of a head appearing on the first toss is $1/2$ and for the head occurring on each successive toss, we have an additional factor of $1/2$, recalling the probability of the multiple futures discussed earlier. In short, the expected winnings can be calculated to be:

$$\frac{1}{2}(\$2) + \frac{1}{2}\left(\frac{1}{2}\right)(\$4) + \frac{1}{2}\left(\frac{1}{2}\right)\left(\frac{1}{2}\right)(\$8) + \cdots$$
$$= \$1 + \$1 + \$1 + \ldots,$$

so that for each additional toss of the coin, we obtain another dollar in the expected value. If the first head does not appear until toss number N, the expected winnings are $\$N$. Consequently, as we play more and longer, the average winnings per game does not converge, but keeps getting larger. This behavior is depicted in Fig. 4.8, where the result of a normal game of chance using an unbiased coin is compared with that of the St. Petersburg game.

The St. Petersburg game leads to a paradox because the banker reasons that the average winnings per game are in theory infinite (or at least very large), thus, the ante should also be large. The bettor, on the other hand, knows that given a large number of games, the bettor usually only wins $1, so s/he wants a small ante. This failure to agree upon the ante is known as the *St. Petersburg paradox*. Literally hundreds of papers have been written on ways to resolve this paradox and thereby make probability theory whole. The paradox is actually a consequence of the scaling property of the game, resulting from the fact that the game has no characteristic scale. We call this *Bernoulli scaling* in which there are two parameters that we can adjust which define the game of chance, one that characterizes

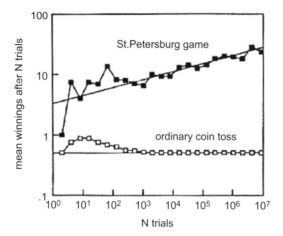

Fig. 4.8: Computer simulation of the average winnings per game after N trails of an ordinary coin toss compared with the St. Petersburg game. The average winnings per game of the ordinary coin toss converge to a finite value. The average winnings per game of the St. Petersburg game continue to increases as it is averaged over an increasing number of games.[47]

the diminishing chances of winning, and the other, the increased return for this unlikely outcome. In general, the power-law index α is determined by the ratio of these two parameters.

In Bernoulli scaling, one parameter is the frequency of occurrence of an event, tossing a given sequence of heads, and the other is the size of the winnings for throwing that sequence. The frequency of occurrence decreases by a factor of two with each additional head, and the size of the winnings increases by a factor of two with each additional head. The increase and decrease are equal and therefore nullify each other and the expected winnings increase in direct proportion to the number of tosses. Bernoulli scaling is observed in the Lévy distribution.

The Lévy distribution, as the solution to the generalized central limit theorem, has a number of interesting properties. The first property of interest is the width of the distribution, which is found to be infinite for various measures. This divergence of the width is not surprising, since this was the very condition that was thrown out in order to generalize the central limit theorem. However, what is the meaning of the width of the distribution being infinite? In order to explain this, I employ the random walk model introduced earlier to help visualize the Gaussian distribution. Recall that a random walk involved a drunk who always takes a step of a given size, but who rotates through a random angle before taking each step. In Fig. 4.9, the random walk is replaced by a random flight, consisting of 10,000 random jumps not all of which are of the same size. In a Lévy-flight, an entity jumps between sites regardless of distance, each in the same time interval. This leads to a divergence in the width of the distribution.

In Fig. 4.9, Klafter *et al.*[48] allow the range of step sizes to change, in particular, to allow the step sizes to be determined by an inverse power

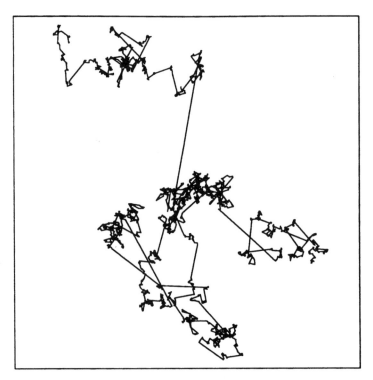

Fig. 4.9: A Lévy flight with a trajectory consisting of 10,000 jumps is shown with the jump distance chosen randomly from an inverse power-law distribution of step sizes with power-law index 2.5. (Taken from Klafter, Zumofen and Shlesinger.[48])

law. In the simple random walk, leading to a Gaussian distribution, all the steps were of the same length and only the rotation angle of the step varied randomly. Bernoulli scaling is one way to generalize the random walk and generate a Lévy-flight. Consider a random flight having the set of jump sizes 1, 10, 100, ... and the associated relative frequencies 1, 0.032, 0.001, ... The unit step has the largest likelihood of occurring, so most of the steps in the walk are of unit length. The next largest step is an order of magnitude larger, but it occurs an order of magnitude less often. For example, a walker may take 10 unit steps on average before taking a larger step, e.g. of length 10 units, and then she continues the unit-step random walking. Thus, on average, a larger step is taken every 10 steps, corresponding to the relative frequency of approximately 0.032. After 10 clusters of 10 unit steps each, on average, the walker may take a step of length 100 units and then continues with the unit-step random walking. The relative frequency of taking a step of length of 100 is 0.001. This process continues to generate progressively longer steps with proportionately lesser frequency.

The structure, within structure, within structure, that is characteristic of a geometrical fractal is replaced with clusters, within clusters, within clusters, for a Lévy flight making the set of points visited during the flight, a statistical fractal. This clumping of the sites visited in the Lévy-flight is quite evident in Fig. 4.9. The simplest form of the Lévy distribution is characterized by a single number named the Lévy index μ, which has a value greater than zero and less than or equal to two. This index can be related to the ratio of the two parameters used to generate the above random walk, the probability of a new scale for the step size being introduced into the random flight and the size of that new scale. Actually, the Lévy index is the ratio of the logarithms of these two parameters, but that level of detail is not for the faint hearted. The clustering property observed in Lévy statistics changes with the value of the Lévy parameter, such that as μ approaches the value two, the clusters merge into one huge blob with the statistics of Gauss.

Lévy distributions in space therefore describe phenomena that statistically cluster in space and such distributions in time have a bursting behavior as determined by the independent variable. These statistics occur in processes where dynamical variables scale and asymptotic values follow an inverse power-law distribution. This is not shown here, but it can be proven that the asymptotic form of the Lévy distribution is an inverse power law, either in space or in time, as shown by Montroll and West.[49] The scaling and clustering properties are actually interrelated, because the properties seen at one scale are repeated at subsequently larger and smaller scales. Consequently, if one observes a statistical separation between events at one resolution, the same statistics can be observed at higher and lower resolutions.

4.6. Only the few matter

Let us look back and see what has been uncovered in this chapter. First of all, I argued that scientists are not the introverted, reclusive scholars with homogeneous personalities invisioned by the public, but they have at least the same level of variability as does the general population. Even my superficial classification of scientific types requires five distinct categories: sleepers, keepers, creepers, leapers and reapers, each of which has its strengths, as well as its weaknesses. The working scientist meanders among these types, staying in some activities longer than others, but usually being driven by the quality of the work being done, if not the recognition gained from doing the work.

Even someone as famous as Jonas Salk, whose contribution to medicine and the alleviation of human suffering is uncontested, did not reap the full rewards of his accomplishments, or at least that was how it appeared to me in the years that I knew him. However, until the end of his life, Jonas was trying to understand human response to diseases. His last medical research activity was collaborating on a vaccine for HIV and not a cure that would erradicate the disease, since he did not think we knew enough to do that. He was working toward the more modest goal of a vaccine that would leave a person symptom-free. He did not succeed but when he died, he was thinking about an interpretation of certain experimental results that might be useful in reaching his goal, not a bad way to die, and, for my money, an even better way to live. One of the many things he did accomplish was to indoctrinate a physicist into some of the ways of the physician.

Most of what was discussed in this chapter concerns the limitations of the bell-shaped curve of Gauss and the linear additive nature of the processes that such distributions describe. Physical phenomena that can be reduced to simple equations of motion (linear) that seem to lend themselves to such descriptions. One variable in the physical system is changed in an experiment and a second variable responds in direct proportion to the change that was made in the first variable. The same thing is true of certain sociological phenomena; we experience this stimulus-response paradigm frequently in our technological world. The harder I step on the brakes, the more rapidly the car slows down. Cars following one another in platoons do not move at a constant velocity; they slow down and speed up in fits and jerks. The fluctuations in velocities in such streams of cars have in fact been found to have Gauss statistics.[2] *Dry Bones* captures the essential linear additve character of such phenomena.

However, these complicated additive phenomena turn out to be the exception and not the rule. Complex phenomena in the social and life sciences are, by and large, not additive; they are instead multiplicative by nature and follow Franklin's lyric rather than the old spiritual. The laws of Pareto, Willis, Auerbach and Zipf, as well as those of Richardson, Lotka and many others,[32] stem from their implicit multiplicative nature which makes them amenable to the *Pareto Principle*. This is the case in which the few are important and determine the properties of complex systems and the many are irrelevant and readily replaceable. More is said about this in the subsequent chapters on network models of complex phenomena and control in such systems.

The *Pareto Principle* was one of the more intriguing properties to emerge from the comparision between the bell-shaped and inverse power-law distributions. However, there is another way to interpret these results, one based on the information concept. Information alone is not a good measure of complexity, since it leads to counterintuitive results, but it is a good measure of the order (regularity) in a time series. The ecological community has introduced a measure of the disorder that estimates the degree to which populations of different species are clumped together or aggregated in space. If I_{max} is the maximum information of a system, then the instantaneous information I can be used to define

$$E = \frac{I}{I_{max}} \qquad (4.2)$$

Entropy

which is the *evenness* of the system. E measures the deviation from equal numbers in the states of the system or the departure from the equiprobable distribution of members. The quantity E is therefore a measure of disorder, with no explicit dependence on the size of the system, having a maximum value of unity and a minimum value of zero. The evenness has its maximum value where the entropy has its maximum value, and its minimum value where the entropy is zero, so that the two are directly proportional.

A measure of order O is defined by subtracting the evenness from one to obtain

$$O = 1 - E. \qquad (4.3)$$

The measure O has a maximum value of unity where the entropy is zero, for example, when the initial configuration of the system is a single, well-defined state. As the system evolves from this state, the entropy increases monotonically towards its maximum value, and the order measure decreases monotonically to zero. A system has maximum disorder when $O = 0$. On the other hand, if entropy alone has been used as the measure of order, there would be disorder that increases with increasing biological complexity, contradicting our understanding of these phenomena.

In information theory, the order parameter defined by Eq. (4.3) is the *redundancy* of the system. The transmission of redundant information, redundancy arising because of the dependence of successive events on one another, lowers the probability of suppressing information through instabilities. Consider a text written in English where we know the English alphabet plus a space contains 27 letters; the amount of information per letter will be $I_{max} = \log_2 27 \approx 4.75$ bits per/letter as we noted earlier. A number of estimates of the actual information content of English text suggest that it is approximately 2 bits per letter. This yields a redundancy of about $O = 1 - 2.0/4.75 \approx 0.58$, or on the average 58% of the information written in English is context specific or redundant. An ecologist might say that the data is 42% clustered. One might conclude that in a dynamical context, it is only the non-redundant information that is valuable; however, it is the redundancy that increases the stability of a system. Therefore, in a nonequilibrium nature, higher levels of organization require a balance between variety and redundancy to generate information while retaining global stability. This may be the essential nature of the 80/20 rule, that the 20% provides for the non-redundant essential information, but the 80% is the redundancy which provides stability for the system.

The quantities E and O are individually not particularly useful measures of complexity. Dynamical systems can be simple (ordered) with only one or a few variables, or with an infinite number of variables. A measure of complexity must therefore vanish at these two extremes, as depicted in Fig. 3.7. Redundancy is central in determining the self-organizing capabilities of a system. For example, the fragile nature of Arctic and boreal fauna is at least partly due to the fact that these ecological systems are not sufficiently redundant to dampen perturbations without becoming unstable. The same is apparently true of large organizations after periods of down-sizing. In the latter situation, redundancy is often interpreted as waste and its removal as efficiency. However, by removing redundancy, the built-in stability of a complex system is sacrificed and inexplicable periods of disorganization often result.

A measure of the delicate balance between redundancy and efficiency in complex systems is given by the simple complexity measure

$$C = EO = E(1 - E) \qquad (4.4)$$

such that complexity C is the product of the measure of disorder E and the measure of order $(1 - E)$. With this compound measure of complexity, as the order goes up, the disorder goes down, each in direct proportion to the other, so that complexity balances the two, and satisfy our intuition. The complexity measure vanishes at total disorder or redundancy ($E = 1$) and at total order ($O = 1$), and is maximum when $E = 1/2$, such that a graph of complexity versus disorder (see Fig. 3.7) has a single maximum of $C = 1/2$ at the eveness value of $E = 1/2$, and monotonically goes to zero on either side of the maximum symmetrically. Note that this measure has a maximum that is not too different from the redundancy of the English language ($O = 0.58$), suggesting that English may be nearly maximally complex using this measure.

Chapter 5

Fractal Physiology

I have argued that the bell-shaped curve, so successful historically in helping to understand experimental variability in the physical sciences, does not translate well into the social and life sciences, despite the best efforts of the leading 19th and 20th century scientists. This failure is not a matter of principle; after all, there is apparently no fundamental reason why the distribution of Gauss should not work in describing medical phenomena. In fact, for 200 years, most scientists became increasingly convinced that the bell-shaped curve should work as well in the social and life sciences as it does in the physical sciences. The problem with this conceptualization of variability and fluctuations is that the data does not support it and the data is the Supreme Court in science; there is no appeal and the experimental decision is final. The inverse power law is the distribution that is ubiquitously observed in the social and life sciences and not the bell-shaped curve of de Moivre and Gauss. I have emphasized some of the different properties in the processes underlying the two kinds of descriptions, but let me now focus more narrowly on the medical phenomena and the data used to characterize them.

In this chapter, I explore what can be learned from the physiological time series data using a number of distinct physiological networks as exemplars: the cardiovascular network, the respiratory network, the motorcontrol network for gait, the regulation of the gut, neural discharge, and finally the thermal regulatory network for body temperature. To simplify things, I put the first five data sets into a common format, that being the time intervals between events. In one case, the event is the maximum contraction during a heartbeat; in another, the event is the maximum expiration during a breath; in the third, the event is the heel strike in a step; for the fourth, the event is a contraction of the stomach; and the fifth is the sharp spike in the action potential. In the final time series, the events are measurements of body temperature at equally-spaced time intervals. If the average were the true indicator of health, then the time intervals would just be a sequence of constants for each phenomenon with any fluctuations being due to measurement error. The time between successive heartbeats would be constant; the time between points of maximum expiration of air from the lungs would be unvarying; the time between successive steps while walking would be fixed; and so forth. However, this is not what has been observed. From the data shown in Figs. 2.1 and 2.6, instead of constant values, the times between events fluctuate for the first three phenomena and the other two will be seen to be the same. Similarly, in a linear, additive world, body temperature would have a fixed value with uncorrelated random fluctuations around this average due to measurement error. However, this is not what is observed either.

The average value is the "gold standard" in medicine, so I should anticipate that the fluctuations in the cardiac time series satisfy a bell-shaped curve, in order for the mean to be the best characterization of the hearbeat time series. A similar expectation should hold for the respiratory and gait time series, as well as for body temperature. If the fluctuation were Gaussian, they would, in fact, be indicative of measurement error and their statistics would hardly contain any interesting information about the physiological phenomenon generating them. The fluctuations might also be biological noise, as many generations of physicians have assumed. However, a number of experiments in the past decade or so have shown that the fluctuations in these time series have inverse power-law rather than Gaussian distributions. In this chapter, I review some recent experimental results and explore the physiological consequences of the inverse power law for the time intervals between successive events.

I now deal with one of those technical issues that requires attention or I could go off in the wrong direction entirely. This issue has to do with the fact that there are two ways inverse power laws can enter into the measurement of physiological phenomena. One way is by means of the statistics of the fluctuations, where the bell-shaped curve is replaced by the inverse power law, as mentioned. This might occur by means of the Gauss distribution being replaced by the Lévy distribution, whose asymptotic form is inverse power-law. The other way, that an inverse power law may enter the data is through long-time memory in the statistics, where the correlation function would be power law, regardless of the form of the statistical distribution. The latter could occur through the process being described as fractional Brownian motion or even as fractional Lévy motion. I discuss these interpretation options more fully after determining the scaling properties of the physiological time series. It is worth emphasizing that the scaling of a fractal time series cannot by itself distinguish among these mechanisms because each is fractal in different ways, as I shall discuss subsequently.

The scaling behavior of fluctuations in experimental time series is measured by means of the power-law index. This empirical index can, in turn, be related to the fractal dimension of the underlying dynamical process, where the fractal dimension for the physiological processes falls between that for a regular (completely predictable) and an uncorrelated random (completely unpredictable) process, that of scaling in *fractal physiology*. Processing the data to determine its scaling properties also needs to be reviewed. This is done using a relatively simple technique from biology involving allometric relations, i.e. power-law relations among system variables.

Scaling does not distinguish between colored noise and intermittent chaos, both of which have inverse power-law spectra.

The picturesque phrase "colored noise" follows Norbert Wiener's study of white light. He knew that white light contains all frequencies in equal spectral weight and random relative phases, consequently, he refered to white light as noise, or more accurately he termed random time series having constant spectral density as white noise. On the other hand, noise that has a non-constant frequency spectrum, that is, noise with a color preference such as inverse power law, was named colored noise to distinguish it from white noise. Not all chaotic time series have an inverse power law spectrum, just those that manifest scaling, which is to say, a dynamic process with bursts, within bursts, within bursts. Therefore, some discussion of forecasting techniques that can discriminate between time series generated by nonlinear dynamical equations and linear stochastic mechanisms with memory, is given. I also indicate how such analyses have been used to support the notion that fractal physiology implies that complex physiologic phenomena are driven by nonlinear dynamical mechanisms.

5.1. Scaling in physiological data

I introduced the term "fractal physiology" in this book as a descriptive metaphor to acknowledge the fact that scaling occurs in all aspects of human physiology. However, this description is much more than a metaphor. Consequently, the view that human physiology occurrs in a Euclidean space of three dimensions is replaced with the notion that physiology occurs in a space of non-integer fractal dimension[50] leading to physiological time series that have fractal statistics. In Chap. 2 we mentioned that the cardiac conduction system, the cardiac wiring that carries the electrical signal from the sino atrial pacemaker cells to the ventricular cells, and which causes the muscles to contract producing the heartbeat, has a ramified branching structure not unlike that of lightning in the night sky. The heart's conduction system has consequently been analyzed in terms of geometrical fractals,[51] but it is not the static anatomical structure that is of most interest here.

Patterns have been sought in the dynamics of the beating heart as well, particularly in the sequence of time intervals between sequential beats. I have argued that the average heartbeat is too restrictive a measure of the health of the cardiovascular system. In fact, such an observation concerning the undesireability of regularity of the heartbeat was first made during the third century A.D. by Wang Shu-He in the Western Chin dynasty[52]:

> If the pattern of the heartbeat becomes regular as the tapping of a woodpecker or the dripping of rain from the roof, the patient will be dead in four days.

A few hardy souls recognized the truth of this observation and knew the phenomenon, if not the quotation. These scientists adopted a complementary perspective and postulated that normal heartbeat has a fractal component in its dynamics, and concluded that chaos is healthy and regularity is pathological[53,54,55] which was subsequently born out by ever-increasing amounts of data.

The medical community is rather conservative. Accepting the notion that fractals and chaos play any role in the normal healthy operation of the cardiovascular system was not embraced with enthusiasm. In fact, when Ary Goldberger and I wrote one of the first papers on the use of fractals and chaos in physiology and medicine, we wanted it published in a mainstream medical journal. However, the paper was rejected by a number of medical journals and it took a few years for this first paper to be accepted for publication.[56] Yet, such resistance to new ideas is not uncommon in science (the keepers are always combing the literature in search of the heretic and on the alert for the insufficiently rigorous). However, there appears to be more resistance to new ideas in medicine than there is in other areas of science. This is probably desirable, as it would not be a good thing for a physician to uncritically accept new ideas and procedures and apply unvetted methods on their patients.

The apparently exotic mathematical concepts of fractals and chaos can be replaced by the single notion of scaling, where the

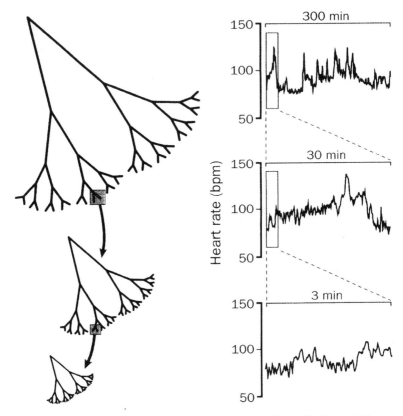

Fig. 5.1: On the left, the scaling embedded in a geometrical fractal is depicted. The arrows indicate a magnification of the boxed region of the fractal bush. On the right, the scaling is embedded in a heart rate time series, which is a statistical fractal, where the magnification of the structure within the box is indicated. In neither case is there a scale characteristic of the underlying structure [taken from Ref. [58] with permission].

concept of scaling and fractal could be loosely used interchangeably, even though this laxness would raise the neck hair of some of my colleagues. Figure 5.1 distinguishes between a geometrical fractal and a statistical fractal. The sketch on the left denotes a sparsely filled "bush" at the top, the kind of structure that Beniot Mandelbrot used so effectively when he introduced the word fractal into the scientific lexicon.[57] The terminal point of one of the branches of the bush, shown as a box, is magnified. This magnified piece of the bush, indicated by the arrow, is identical to the original bush, only smaller in overall size (a matter of scale). Again, if a terminal point of one of the branches of this magnified bush is magnified, the twice-magnified piece is identical to but smaller than the original bush. The magnified parts of the bush are not made the same size as the original in order to visually emphasize that this is embedded within the larger structure. Thus, the structure of the original bush repeats itself on smaller and smaller scales, with no single scale being characteristic of the structure. This property of having scale-free, exact reproduction of the structure at smaller and smaller scales defines a geometrical fractal.

On the right-hand side of Fig. 5.1 is a data set, the heart's beat-to-beat intervals of a healthy young male. These data are depicted at three different time resolutions. The first time series is 300 minutes (5 hours) long and looks erratic. The boxed region at the top of the figure is a 30 minute segment of the original time series and is magnified in the middle figure to fill the same horizontal distance as the original data set. Although the magnified segment of the time series is not identical with the original time series, as it was with the geometrical fractal, it looks to have approximately the same number of bumps and wiggles. Focusing on the boxed region of this second time series, for a segment of 3 minutes long, let us stretch this segment out to be the same length as the other two. Again, although we have changed the time scale by a factor of one hundred from the original to the final time series, the variability does not look significantly different from the other two. It is not the exact geometrical structure that is reproducing itself, but rather the statistical structure that is being reproduced or scaled. Thus, time series of this kind are known as statistical fractals and the corresponding probability density has the scaling property and not the stochastic process itself.

The pattern that is identified on the right-hand side of Fig. 5.1 has to do with the variability of the time interval from one heartbeat to the next. The pattern is in terms of the deviation of the heart rate from its average value and is somewhat elusive. In the cardiac literature, this deviation in the heart rate is called heart rate variability (HRV) and its statistical distribution has a number of interesting properties. One property that should resonate with this discussion is that the statistics for the inter-beat intervals satisfy an inverse power-law distribution. This, however, is only part of the story.

From this discussion, scaling appears to be a relatively new idea, although it is not. In 1753, Jonathan Swift wrote:

> So, Nat'ralists observe, a Flea
> Hath smaller Fleas that on him prey,
> And these have smaller Fleas to bit'em
> And so proceed ad infinitum

These couplets captured an essential feature of what was then natural philosophy; the notion that the dynamical activity we observe in natural phenomena is a consequence of many unseen layers of movement and that this imperceptible motion is related from one level to the next by means of scaling. Swift was reflecting on self-similarity between scales, i.e. smaller versions of what is observed on the largest scale repeats in a decreasing cascade of activity to smaller and smaller scales. Of course, this is not the only way fractals are formed, scaling downwards to ever-smaller structures; they can also blossom upwards as anticipated by de Morgan in 1872,

> Great fleas have little fleas
> upon their backs to bite'em
> And little fleas have lesser fleas,
> and so ad infinitum.
> And the great fleas themselves,
> in turn, have greater fleas to go on,
> While these again have greater still,
> and greater still, and so on.

where the poet captures an essential feature of nature as recorded in Mandelbrot's book on fractals.[57]

The mathematical notion of a fractal does not have a simple definition, but all attempts at one incorporates the above idea that the whole is made up of parts similar to the whole in some way. This intuitive idea is consistent with the self-similar shrinking downwards or expanding upwards observed by the poets. In a mathematical fractal, the proliferating structure proceeds away from one's scale of reference in both directions without limit. In a natural fractal, one invariably encounters the largest and the smallest scales because the mechanism controlling the system changes at the extremes, at the limits of the very small and very large. If the fractal concept is to be useful, there must be a substantial domain that is scale-free and it is over this region that the natural fractal mimics the mathematical fractal. So how can these ideas be incorporated into our understanding of the interbeat interval time series?

A geometrical fractal has no fundamental scale, so identical geometric structure is repeated on smaller and smaller scales. Suppose I want to know the length of one of the branches of the bush in Fig. 5.1. My first assumption is that the length of a branch is finite and therefore it can be measured. This assumption of a finite length in a finite interval is blatantly obvious which is worth investigating, especially when it turns out to be false. Suppose I measured the length of the branch on the top bush using a ruler no shorter than the shortest segment I can see. I would surely measure a finite value for the total length of a branch. By magnifying the tip of the bush, a lot of details becomes visible on this new scale of resolution that were missing from my first measurement, because I excluded this small-scale structure. I therefore choose a smaller ruler of the size of the tip in the magnified segment, and measure the total length of the branch again, this time including the small-scale structure. The length is now longer than that found in the first measurement. Where this is heading is evident since the tip of the magnified branch can be magnified again, another measurement can be made with an even smaller ruler and the measured length of the branch will again increase due to the shorter length of the new ruler. This argument reveals that the measured length of the branch gets increasingly longer as the size of the ruler gets increasingly shorter, and consequently includes more structure in the measurement. Thus, the length of a fractal branch depends on the size of the ruler used to measure it and as the size of the ruler becomes infinitesimally small, the length of the branch becomes infinitely long. The infinite length occurs because there is no scale at which the fractal object becomes smooth. (Recall the discussion in the section on non-traditional truths regarding the lack of scale in the fractal objects.)

A statistical fractal, of the kind generated by the generalized central limit theorem, has this fractal repetition of structure from scale to scale as well. What this means in practice is that, just as in the case of the geometrical fractal, what I observe in a statistical fractal depends on the resolution of the instrument I use to make the observation. For example, a time series with such fractal behavior in time would have a burst of activity followed by a quiescent period, followed by another burst of activity. However, if I magnified the region of the burst, the activity would not fill the interval uniformly. Instead, I would see a smaller burst of activity followed by a quiescent period and then another smaller burst of activity. This is the phenomenon of clusters within clusters within clusters, which I described in the discussion of Lévy flights, but now the clusters are in time rather than space, the so-called *fractal time*. In medicine, such time series are usually smoothed on these smaller time scales, because it is believed that what occurs at these small scales is just noise and carries no information about the physiological network being studied. However, there can be a great deal of information in these fluctuations, depending on whether the fluctuations are generated by chaotic dynamics or by noise. This is the information that is lost by using average values such as the average heart rate. Furthermore, there is now a way to test the truth of the hoary statistical assumption that smoothing over small-scale fluctuations in time series is a reasonable, but more importantly, a useful thing to do.

5.2. Allometric relationships

L.R.Taylor[59] was a scientist interested in biological speciation, such as how many species of beetle can be found on a given plot of land.

He answered this question by sectioning off a large field into parcels and in each parcel sampling, the soil for the variety of beetles that were present. This enabled him to determine the distribution in the number of new species of beetle spatially distributed across the field. From the distribution, he could then determine the average number of species denoted by the symbol μ and the standard deviation around this average value, denoted by the symbol σ. After this first calculation, he partitioned his field into smaller parcels and redid his examination, again determining the mean and standard deviation in the number of species at this increased resolution. This process was repeated a number of times, yielding a set of values of μ and σ. In the ecological literature, a graph of the logarithm of the variance σ^2 versus the logarithm of the average μ is called a power curve.[60] The algebraic form of the relation between the variance and average is

$$\sigma^2 = a\mu^b, \qquad (5.1)$$

where the two parameters a and b determine how the variance and mean are related to one another.

A hypothetical data set satisfying Eq. (5.1) is plotted in Fig. 5.2(a). The variance on the vertical axis grows nonlinearly with the increasing average on the horizontal axis, with some fluctuations such as might be observed in real data sets. If the data in Fig. 5.2(a) do satisfy the power-law relation of Taylor, then I can take the logarithm of the data to obtain the graph shown in Fig. 5.2(b). Again in the graph of the logarithm of the variance versus the logarithm of the average I see some fluctuations, but the dominant behavior is a direct proportionality between the two logarithmically transformed variables. The proportionality constant is the slope of the curve, which algebraically I can identify with the power-law index in equation (5.1). A plot of this relation on doubly logarithmic graph paper yields a straight line with slope given by b and intercept given by $log\ a$. In Fig. 5.3, we sketch the logarithm of the variance versus the logarithm of the average for three typical values of the scaling exponent.

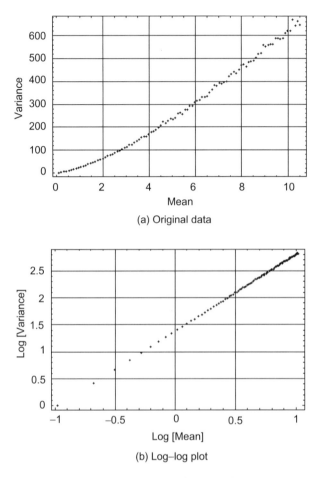

Fig. 5.2: (a) Computer-generated data that satisfies Taylor's power-law relation between the variance and mean is plotted. (b) The same data used in (a), after taking the logarithm of the variance and the logarithm of the mean, is plotted. It is evident that the log-log graph yields a straight line when the variance and average satisfy the Taylor's Eq. (5.1).

Taylor was able to exploit the curves in Fig. 5.3 in a number of ways using the two parameters. If the slope of the curve and the intercept are both equal to one, $a = b = 1$, then the variance and the mean are equal. This equality is only true for a Poisson distribution, which, when it occurred, allowed him to interpret the number of new species as being randomly distributed over the field with the number

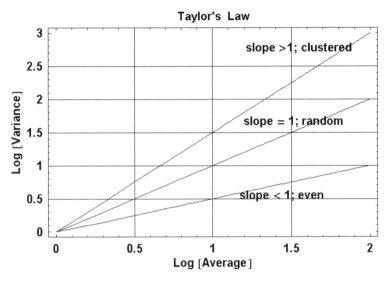

Fig. 5.3: The power curve is a graph of the logarithm of the variance versus the logarithm of the average value. Here, three such power curves are depicted in which the three ranges of values for the slope determine the general properties of the underlying statistical distribution of new species.

of species in any one parcel being completely independent of the number of species in any other parcels. However, if the slope of the curve was less than unity, the number of new species appearing in the parcels was interpreted to be quite regular. The spatial regularity of the number of species was compared with the trees in an orchard and given the name *evenness*. We encountered this concept of evenness using information ratios and the fact that it was used as a measure of the loss of clumpiness. Finally, if the slope was greater than one, the number of new species was clustered in space, like disjoint herds of sheep grazing in a meadow, might be the equivalent of a Lévy flight.

Of particular interest to us was the mechanism that Taylor and Taylor[61] postulated to account for the experimentally observed allometric relation:

> We would argue that all spatial dispostions can legitimately be regarded as resulting from the balance between two fundamental antithetical sets of behavior always present between individuals.

These are repulsion behaviour which result from the selection pressure for individuals to maximize their resources and hence to separate, as well as attraction behavior, which results from the selection pressure to make the maximum use of available resources and hence to congregate wherever these resources are currently most abundant.

Consequently, they postulated that it is the balance between the attraction and repulsion, migration and congregation, that produces the interdependence of the spatial variance and the average population density.

I can now interpret Taylor's observations more completely because the kind of clustering he observed in the spatial distribution of species number, when the slope of the power curve is greater than one, is consistent with an asymptotic inverse power-law distribution. Furthermore, the clustering or clumping of events is due to the fractal nature of the underlying dynamics. Willis,[34] some 40 years before Taylor, established the inverse power-law form of the number of species belonging to a given genre as mentioned in the previous chapter. Willis used an argument associating the number of species with the size of the area they inhabited. It was not until the decade of the 1990s that it became clear to more than a handful of experts that the relationship between an underlying fractal process and its space filling character obeys a scaling law. It is this scaling law that is reflected in the *allometric relation* between the variance and the average.

Allometric relations are not new to science, they were introduced into biology in the 19th century. Typically, an allometric relation interrelates two properties of a given organism. For example, the total mass of a deer is proportional to the mass of the deer's antlers raised to a power. Thus, on a doubly logarithmic graph, this relation would yield a straight line with a slope given by the power-law index. Huxley summarized the experimental basis for this relation in his 1931 book, *Problems of Relative Growth*,[62] and developed the mathematics to describe and explain allometric growth laws. In physiological systems, he reasoned that two parts of an organism grow at different rates, even though the rates are proportional to one another. Consequently, how rapidly one part of the organism grows can be related to how rapidly the other part of the organism grows. In fact, Taylor's Law relating the variance to the average value is just another kind of allometric relation in which the central moments properties of the statistics are tied together.

It is possible to test the allometric relation of Taylor using computer-generated data. However, we also note that Taylor and Woiwod[63] were able to extend the discussion from the stability of the population density in space, independent of time, to the stability of the population density in time, independent of space. Consequently, just as spatial stability, as measured by the variance, is a power function of the mean population density over a given area at all times, so too is the temporal stability measured by the variance, a power function of the mean population density over time at all locations. With this generalization in hand, I can apply Taylor's ideas to time series.

At the top of Fig. 5.4 is displayed 512 data points generated from a computer package implementing Gaussian statistics, having a mean of one and a standard deviation of two. The top figure displays all the data points generated by the package. The middle figure shows the time series with every four adjacent ($m = 4$) data points added together, yielding a total of 128 aggregated data points. The lower figure shows the time series with the nearest eight data points averaged together, yielding a total of 64 aggregated data points. What is being accomplished by aggregating the data points in this way is evident from the figure. The standard deviation is reduced and the mean is shifted, as more and more of the data points are aggregated.

Figure 5.5 graphs the logarithm of the variance versus the logarithm of the average, for each level of aggregation of the data shown in Fig. 5.4. At the extreme left, the first dot denotes the value of the variance and average obtained using all the data. Moving from left to right, the next dot we encounter corresponds to the variance and average for the time series with two adjacent elements added together. Each of the successive dots, moving from left to right, correspond

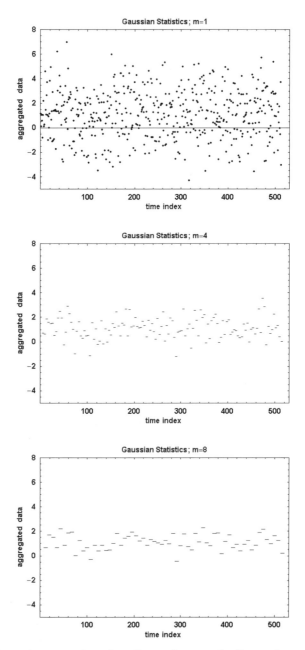

Fig. 5.4: Successive aggregating of *m* adjacent elements of a discrete time series with an uncorrelated Gaussian statistical distribution is depicted.

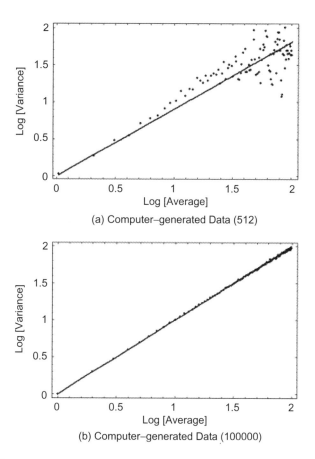

Fig. 5.5: The computer-generated data depicted in Fig. 5.4 is used to calculate the variance and average value for each level of aggregation in (a). In (b), the same procedure is followed for one hundred thousand computer-generated data points.

to the ever-greater aggregation numbers. Consequently, this process of aggregating the data is equivalent to decreasing the resolution of the time series, and as the resolution is systematically decreased, the adopted measure, the relationship between the mean and the variance, reveals an underlying property of the time series that was present all along. The increase in the variance with increasing average for increasing aggregation number shown in the figure is not an arbitrary pattern. The relationship indicates that the aggregated uncorrelated data points are interconnected. The original data points are not correlated, but the addition of data in the aggregation process induces a correlation, one that is completely predictable. The induced correlation is linear if the original data is uncorrelated, but not linear if the original data is correlated. This is the kind of property that Mark Kac would have loved, i.e. data that are linearly independent but not completely independent.

The solid line in Fig. 5.5 is the "best" straight line that can be drawn through all the data points. This line is the best linear indicator of how the variance is related to the average value. If I were given to writing equations, I would see that the logarithm of the variance is directly proportional to the logarithm of the average value, that is, the two logarithms are proportional to one another; as one changes so does the other, in direct proportion to that change. If the relationship is not depicted in the figure in terms of logarithms, but rather, it is expressed in terms of the original variables, the variance and the average value, the allometric (power-law) relation in Eq. (5.1) would result. The process of addition introduces a specific kind of correlation, one in which the slope is one when the initial time series is an uncorrelated random process. However, deviations of the scaling parameter from one indicates that the underlying time series was in fact correlated.

I am interested in the measure provided by the power-law index. I can relate the fractal dimension D for the time series to the slope of the variance versus average value curve in Fig. 5.5, as follows, $D = 2 - slope/2$. In the example of 512 data points, the slope is positive with a value of 0.82, so that the fractal dimension of the time series is $D = 1.59$. However, there is a great deal of scatter of the data points around the straight line yielding the slope as the number of aggregated points gets larger. Note that for an infinitely long data set with uncorrelated Gaussian statistics, the fractal dimension should be 1.5, and that would have been the number obtained if the computer-generated time series were infinitely long. This latter case is depicted in Fig. 5.5(b) where I have done the same processing on a hundred thousand computer-generated uncorrelated Gaussian data points. Notice how much more smoothly the data points fall along the straight line with the unit slope.

Doing a computer calculation of a statistical time series is like doing an experiment, you never get the same answer twice; the uniqueness arises because each time series is finite, so the exact theoretical value is never obtained. The difference between the theoretical value of the fractal dimension $D = 1.5$ and the one obtained for the finite data set generated here is due to the small number of data points (512). In a realization with such a small number of data points, a random string of a few similar values could be misinterpreted in the processing to be a correlation or anti-correlation, and consequently result in a modified fractal dimension. This modification is lost as the number of data points increases, since any finite patch of data becomes proportionately less significant as the total number of data points increases. Readers may not be familiar with the term "anti-correlation". When two quantities are positively correlated, they move together such that when one increases/decreases, the other increases/decreases as well. When two quantities are negatively correlated, they move in opposite directions, such that when one increases/decreases, the other decreases/increases. The former quantities are said to be correlated, whereas the latter quantities are said to be anti-correlated.

5.3. Fractal heartbeats

The mechanisms producing the observed variability in the size of the heart's interbeat intervals apparently arise from a number of sources. The sinus node (the heart's natural pacemaker) receives signals from the autonomic (involuntary) portion of the nervous system which has two major branches: the parasympathetic, whose stimulation decreases the firing rate of the sinus node, and the sympathetic, whose stimulation increases the firing rate of the sinus node pacemaker cells. The influence of these two branches produces a continual tug-of-war on the sinus node, one decreasing and the other increasing the heart rate. It has been suggested that it is this tug-of-war that produces the fluctuations in the heart rate of healthy subjects, but alternate suggestions will be pursued subsequently. Consequently, heart rate variability (HRV) provides a window through which we can observe the heart's ability to respond to normal disturbances that can affect its rhythm. The clinician focuses on retaining the balance in regulatory impulses from the vagus nerve and sympathetic nervous system, and in this effort, it requires a robust measure of that balance. A quantitative measure of HRV serves this purpose.

Heartbeat variability or heart rate variability are both denoted by HRV, since they refer to inverse measures of the same process; a rate being the inverse of a time interval. HRV has become very well known over the past two decades as a quantitative measure of autonomic activity. The medical community became interested in developing such a measure of heart rate variability because experiments indicated a relationship between lethal arrhythmias and such activity, which is discussed in subsequent chapters. The importance of HRV to medicine became widely apparent when a task force was formed by the *Board of the European Society of Cardiology* and the *North American Society of Pacing and Electrophysiology*, and was tasked with the responsibility of developing the standards of measurement, physiological interpretation and clinical use of HRV. The task force published their findings in 1996.[64] It is one of the rare occasions that the members of such a task force were drawn from the fields of mathematics, engineering, physiology and clinical medicine in recognition of the complexity of the phenomenon they were tasked to investigate and actually worked together.

A heartbeat of 60 bpm literally means that the heart beats 60 times every minute. This is simple to understand and simple to communicate to a patient. If 60 bpm is normal for a person at rest, then it is clear that at 30 bpm, the heart is not working as hard as it should and ought to be stimulated. A rate of 120 bpm for that person at rest means that the heart is running wild and must be brought under control. Either extreme is readily identified and the corresponding remedy is relatively straightforward to articulate, if not to achieve. Of course, the natural heart rate varies from person to person, so that an individual with a normal heart rate of 40 bpm would not be alarmed with a single measurement of 30 bpm. On the other hand, when a person with a normal heart rate of 80 bpm sees a measurement of 30 bpm, he or she should be quite concerned and see a physician immediately.

When the heart is doing something out of the ordinary, it becomes very important how to quantify the variation in heartbeat or heart rate. The degree of deviation is determined by the interpretation of the size of the variation and how it is used to identify associated patterns. There are approximately 16 ways to assess HRV in literature, each related to scaling in one way or another. I have adopted a relatively straightforward simple explanatory technique which allows me to introduce the measure of interest. The procedure is closely related to what I did using pictures to explain the scaling properties of fractals in Fig. 5.1. However, instead of zooming in on a point in a time series and increasing the resolution, the resolution was made worse. The resolution was decreased because fractals ought to scale whether or not I increase or decrease the resolution, and decreasing the resolution is much easier with real data. In addition, I can use the allometric aggregation technique on real data to relate the variance and average, as discussed in the previous section.

One way to measure the interbeat interval time series of the human heart is to record the electrical signal at the surface of the skin produced by the heart's pulsation and the pumping of blood. The computer stores the electronic analogue of the strip chart recording of an ECG, which is a sequence of QRST pulses, as shown in Fig. 2.5. An algorithm can then determine the local highest point in the analogue signal and record the time at which this maximum occurs. The difference between the times of successive maxima yields a sequence of the time intervals from one maximum to the next, which is the RR-interval time series. How well this procedure does is determined by the accuracy with which the true maxima of the R-component of a pulse are located. The computer does not record the continuous trace seen on the strip chart. Instead, it samples this curve once every fraction of a second. For instance, a sensor samples the signal at 100 Hz and records it on a computer. This means that the computer records the value of the signal one hundred times a second. Consequently, no computer algorithm can determine the maximum value of the signal to a greater resolution than 0.01 seconds. However, scientists typically assume that the signal is continuous between measurements and they program this assumption into the computer to process data. This assumption is probably good enough for most applications, but in some cases, this resolution may be a few percent of the signal; the signal being the sequence of time intervals between successive heartbeats.

To increase the resolution of the time series might require constructing a new electronic circuit to sample the analogue signal at a higher rate, or more practically, it might require buying such a circuit from a vendor. However, for some technical applications, it is not always possible to obtain the piece of equipment needed commercially and must be built instead. Thus, increasing the resolution in an experiment is not always an option, but it is always a problem. Therefore, I do not want to increase the resolution of a given time series. Instead, I choose to circumvent this problem by decreaseing the resolution of the time series. I demonstrated how to decrease the resolution using a computer-generated data set with known statistical properties in the discussion of Taylor's Law.

I now apply the allometric aggregation approach to the RR-intervals shown in Fig. 5.6, where a typical HRV time series for a healthy young adult male is traced. A time series that has the same number of data points as in the computer-generated time series shown in Fig. 5.4(a) is depicted. The data points in the figure are connected to aid in visualizing how the time intervals between heartbeats are changing. It is evident that the variation in the time intervals

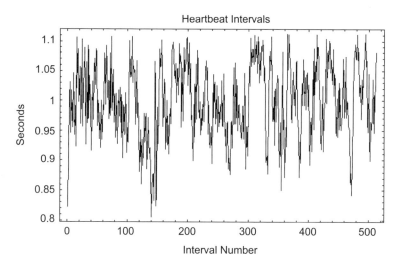

Fig. 5.6: The time series of heartbeat intervals of a healthy young adult male is shown. It is clear that the variation in the time interval between beats is relatively modest, but certainly not negligible.

between heartbeats is relatively small, the mean being 1.0 sec and the standard deviation being 0.06 sec. This relatively modest variance apparently supports the frequently used medical term "normal sinus rhythm". So what do we learn by applying the allometric aggregation technique to these data and constructing the standard deviation and average as a function of aggregation number?

In Fig. 5.7, the logarithm of the aggregated standard deviation is plotted versus the logarithm of the aggregated average value for the HRV time series depicted in Fig. 5.6. Note that the standard deviation is displayed in the figure and not the variance used in the discussion of Taylor's Law. I use the standard deviation because I am primarily interested in whether the time series is fractal and not particularly in the actual value of the fractal dimension. At the left-most position, the data point indicates the standard deviation and average, using all the data points. Moving from left to right, the next data point is constructed from the time series with two nearest-neighbor data points added together and the procedure is repeated moving right,

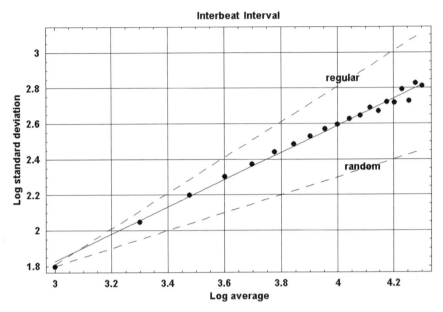

Fig. 5.7: The logarithm of the standard deviation is plotted versus the logarithm of the average value for the interbeat interval time series for a young adult male using different values of the aggregation number. The solid line is the best fit to the data points and yields a fractal dimension of $D = 1.24$ midway between the regular curve and the uncorrelated random curve.

until the right-most data point has 20 nearest-neighbor data points added together. The solid line is the best linear representation of the scaling and intercepts most of the data points with a positive slope of 0.76. We can see that the slope of the HRV data is midway between the dashed curves depicting an uncorrelated random process (slope $= 1/2$) and one that is deterministically regular (slope $= 1$).

The first conclusion drawn is that even though the exact value of the fractal dimension obtained here may not be reliable because of the shortness of the time series, the interbeat intervals do not form an uncorrelated random sequence. I should emphasize that the conclusions drawn are not from this single figure or set of data alone, these are only representative of a much larger body of work. The conclusions are based on a large number of similar observations made by myself and colleagues using a variety of data processing techniques, all of which yield results consistent with the scaling of the HRV time series (e.g. see Ref. 65).

Phenomena obeying a scaling relation, such as shown for the HRV time series data in Fig. 5.7 are said to be self-similar. The fact that the standard deviation and the average value change as a function of aggregation number, implies that the magnitudes of the standard deviation and average value depend on the size of the ruler used to measure the time interval. Recall that this was one of the defining characteristics of a fractal curve, where the length of the curve becomes infinite as the size of the ruler goes to zero. The dependence of the average and standard deviation on the ruler size, for a self-similar time series, implies that the statistical process is fractal and consequently defines a fractal dimension for the HRV time series. Note, I do not ask the reader to accept the conclusion that the HRV time series is fractal, based on the single example given here, but refer you to the literature which has been steadily growing since I began collaborations in this area in the early 1980s (e.g. the reviews in *Physiology, Promiscuity and Prophecy at the Millennium: A Tale of Tails*[32] and Suki et al.[66]).

It is interesting to recall the mechanism postulated by Taylor and Woiwod to explain the allometric relation. They reasoned that the power-law relation results from the balance between two competing processes, migration and congregation, for the case of individuals distributed in space. For the HRV time series discussed in this section, there is an analogous mechanism resulting from the competition between the sympathetic and parasympathetic nervous systems. The continuing tug-of-war results in the density-dependent variance and mean manifested in the power-law relation for the beat-to-beat intervals.

The average scaling exponent obtained by Peng et al.[67] for a group of 10 healthy subjects having a mean age of 44 years, using 10,000 data points, was $\alpha = 0.19$ for the difference in beat intervals time series. Consequently, healthy subjects present scale-invariant, long-range anti-correlations. They[67] interpreted this value to be consistent with a theoretical value of $\alpha = 0$, which they conjectured would be obtained for an infinitely long-time series. The latter scaling implies that the scaling exponent for the beat intervals themselves would be 1.0. However, all data sets are finite and it is determined that the asymptotic scaling coefficients for the interbeat interval time series of healthy young adults lie in the interval [0.7,1.0].[68,69] The value of the scaling coefficient obtained using much shorter time series and a relatively simple processing technique is consistent with these results.

I should also comment on certain speculations made by Peng et al.[67] regarding their analysis of the set of HRV time series; speculations that have been supported by subsequent research. They proposed that the scaling behavior is adaptive for two reasons:

(i) The long-range correlations serve as an organizing principle for highly complex, nonlinear processes that generate fluctuations on a wide range of time scales.
(ii) The lack of a characteristic scale helps prevent excessive mode-locking that would restrict the functional responsiveness of the organism.

Unsurprisingly, these conjectures were the natural result of the application of the notion of fractal physiology to the heart, since Goldberger, a co-author of the above paper, was a collaborator on the development of both ideas. In the following chapter, we incorporate these conjectures into a new perspective on the body's control of physiological systems.

5.4. Intermittent chaos and colored noise

The evidence presented up to this point has established that HRV time series do scale. However, scaling alone does not distinguish between nonlinear dynamic and colored noise processes. An intermittent chaotic process, generated by nonlinear dynamics, has an inverse power-law spectrum. On the other hand, colored noise generated by fractional Brownian motion is also a fractal stochastic process. In fact, the HRV models discussed by many authors use various generators of colored noise to explain the fractal or scaling nature of HRV time series, while still suggesting that the underlying mechanism might in fact be chaotic.

The two phenomena, chaos and colored noise, ought to be distinguishable, if only investigators were sufficiently clever. Since there is a deterministic generator of chaos, one ought to be able to predict its evolution, at least for short intervals of time. In fact, even beyond where one can faithfully forecast a chaotic signal, the deviation of a forecast from the data (the error) behaves in a predictable way. Therefore, one can use the growth of error in a forecast to determine if a given time series is chaos or noise, since the two kinds of error grow differently. Most forecasting is currently done with linear models, and since linear dynamics cannot produce chaos, these methods cannot produce good forecasts for chaotic time series. Although the development of nonlinear forecasting techniques is an active area of research, with few exceptions, the full exploitation of the concept of chaos has not yet been made. However, investigations have been done to compare the quality of forecasts using linear and nonlinear techniques of HRV time series with some success.

When the time series being investigated is chaotic, the system dynamics may be modeled by: (i) finding a state space with minimum embedding dimension, that is, a state space having the absolute miminum number of variables necessary to describe the system's dynamics, and (ii) fitting a nonlinear function to a map that transforms the present state of a system into its future states. In this way, a time series can be used to determine not only the number of variables necessary to model the underlying dynamical system, but the data stream can also provide an approximate functional form for the nonlinear dynamics governing the evolution of the system. The great promise of chaos is that although it puts limits on long-term predictions, it implies predictability over the short term. Nonetheless, how well this can be done in a medical context remains to be seen.

One way to find the state space in which the dynamics of the phenomenon of interest unfolds over time, is to use what is known as the attractor reconstruction technique (ART). This is accomplished by segmenting the data record into a number of pieces and identifying each piece with a distinct dynamical variable. If three such partitionings of the data captures the essential behavior of the system, then only three variables are required to describe the dynamics. If four partitionings are required, then four variables are necessary, and so forth. If the attractor in the phase space in which the dynamics takes place has a dimension d, then a necessary condition for determinism is that the number of variables is greater than or equal to d.

For example, suppose the location of the bob of a swinging pendulum is recorded every 0.1 second, this sequence of locations, at equally spaced time intervals, could be used to construct the position and the velocity of the pendulum as a point in the appropriate phase space, (see Fig. 5.8). A closed orbit traces out an ellipse in phase space where each of the closed curves corresponds to a different total energy, with the larger orbits having the higher energy. For low energies, this system is linear, harmonic and exactly predictable; at higher energies, the system is nonlinear and periodic, but still exactly predictable. However, at the separatrix where the closed swinging motion of the pendulum is separated from its rotating motion about the pivot, strange things can happen. This is where chaos can occur and ART can distinguish between the three kinds of motion: bounded, unbounded and chaotic.

Once the phase space is found using ART, a model can be constructed to fit the data. This model has the form of a transformation that maps the data presently available into the future; in other words, it can be used to make a forecast. If I only use part of the data to construct the transformation to make the forecast, then the remainder of the data can be used to test the quality of the forecast. The difference between the forecast and the actual data defines an error, and since

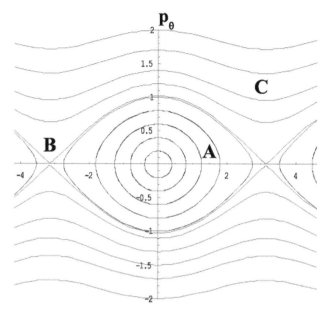

Fig. 5.8: The phase space for the equal energy surface of a pendulum are depicted. The closed curves A denote bounded motion, the swinging of the pendulum from side to side. The open curves C denote libration, or the rotation of the pendulum around the pivot point. The curve that separates the two types of motion B is the separatrix where the pendulum can be delicately balanced between the two types of motion. It is the separatrix that can characterize chaotic behavior.

the data set of interest is fluctuating, I focus on the average error, that is, the average difference between the actual and predicted data divided by the standard deviation of the data. If the average error is zero, the forecast is "exact". If the average error is one, the forecast is no better than it could have been done using the average value of the data to make the forecast. Hence, the importance of variability over the average value is clear in determining the behavior of the nonlinear system.

A well-behaved mapping function or transformation can be easily modeled using any of the numerous representations, provided the data stream is sufficiently long. For a complicated function, one having many variations in a short time interval, it is unclear that any representation can provide an adequate approximation to the actual dynamics. The dependence on representation can be reduced by means of a local approximation in which the domain of the data is segmented into local neighborhoods.[70] Using this approach, the growth of error with the length of the prediction time is very different using linear as opposed to nonlinear forecasting functions. Consequently, to determine if a time series has a strong nonlinear component in its dynamics, error can be defined as the difference between the two types of predictions made using a linear and using a nonlinear mapping function. A strongly divergent error between the two predictions indicates that the linear and nonlinear transforms yield very different forecasts, implying that the underlying time series is nonlinear, otherwise, the two predictions would be essentially equivalent.

What would be the result of applying this forecasting technique to colored noise? A linear stochastic model, such as a moving average or an autoregressive model, are usually adequate to generate colored noise, so one should see no improvement in the above forecasting method using a nonlinear rather than a linear forecasting transform. Consequently, an increasing error between the linear and nonlinear forecasts implies that the underlying cause of the variability in the experimental time series is due to a nonlinear dynamic process and consequently, it is not colored noise.

An example of how to apply ART is depicted in Fig. 5.9 using heart interval data. The HRV data is denoted as Data(t) which is plotted versus the same time series shifted by one heartbeat and denoted as Data(t+1). In the upper figure, a long narrow sequence of loops is shown to lie within the data and this structure might be interpreted as an attractor on which the cardiac dynamics unfolds. An argument for this interpretation is given by the lower figure where the same data are used, but their ordering in the time sequence has been randomly shuffled. It is evident that randomly shuffling the data points results in the loss of the phase space structure, giving rise to a Gaussian blob of data. The structure seen in the upper figure is a consequence of the time ordering of the data points. This order is a consequence of

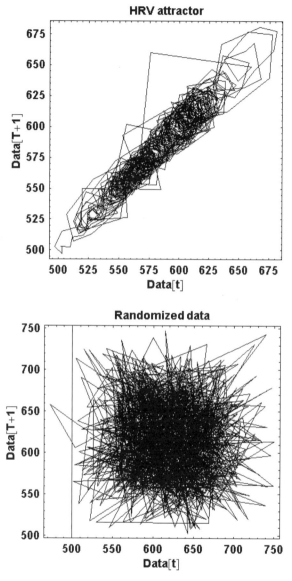

Fig. 5.9: The upper graph is the HRV data (1000 heart beats) plotted versus a copy of itself shifted in time by one hearbeat. The lower graph is the same plot, but with the ordering of the data points randomized.

the linear and nonlinear correlation of the data as evidenced by the loss of that structure when that time ordering is disrupted, as in the lower figure.

Sugihara et al.[71] applied a form of the ART method to the nonlinear analysis of infant heart rhythms and observed a marked rise in the complexity of the electrocardiogram with maturation:

> ...that normal mature infants (gestation $\geq$ 35 weeks) have complex and distinctly nonlinear heart rhythms (consistent with recent reports of healthy adults)....These three lines of evidence support the hypothesis championed by Goldberger et al. [Goldberger, A.L., Rigney, D.R. & West, B.J. (1990) Sci Am **262**: 43–49] that autonomic nervous system contol underlies the nonlinearity and possible chaos of normal heart rhythms. This report demonstrates the acquistion of nonlinear heart rate dynamics and possible chaos in developing human infants...

I return to this paper in the chapter on the nature of disease. At this moment, I emphasize that the analysis of Sugihara et al. was the first quantitative analysis of HRV data that directly supported the conjecture of the heartbeat being made chaotic by Goldberger and West in 1987.[56]

5.5. Fractal breathing

Breathing and respiration are often used interchangeably, which may be adequate for informal conversation, but not for serious discussion, because there are real differences between the two. Breathing is a function of the lungs whereby the body takes in oxygen and expels carbon dioxide, whereas respiration is the energy-producing process by which the oxygen taken in by the lungs is distributed evenly in the body, used for oxidation of food, and gives off the carbon dioxide expelled by the lungs. The chemistry of the energy releasing process involves the consumption of oxygen by cells that need it, in order for oxygen to react with glucose and make carbon dioxide, water, and energy in the form of ATP. The two terms, breathing and respiration, are essentially equivalent when one is considering simple breathing; otherwise, the more technical term of respiration would be more appropriate.

Like the heart, the smooth muscles in the bronchial tree are innervated by sympathetic and parasympathetic fibers, and produces contractions in response to stimuli such as increased carbon dioxide, decreased oxygen and deflation of the lungs.[72] Fresh air is transported through some 20 generations of bifurcating airways of the lungs, during inspiration, down to the alveoli in the last four generations of the bronchial tree. At this tiny scale, there is a rich capillary network that interfaces with the lungs for the purpose of exchanging gases with the blood.

The bronchial tree has a kind of static or structural complexity, which is different from the dynamic complexity we focused on in the previous section. The lung may indeed serve as a useful paradigm for anatomic complexity because of its branching tree structure. One can see this tree-like network of a complicated hierarchy of airways, beginning with the trachea and branching down to an increasingly smaller scale of the level of tiny tubes called bronchioles, ending in the alveoli sacs. What is truly remarkable is that the surface area of the alveolar, where the gas exchange actually takes place in adults, totals 50 to 100 square meters, approximately the size of a standard tennis court. These surface areas are crinkled up to fit onto the surface of the lung in the chest cavity, the distal connections of the bronchial tree.

In this section, I am not only interested in the architecture of the mammalian lung, but also in the dynamics of breathing; the apparently regular breathing that take place as you sit quietly reading this book. Here, evolution's development of the lung over millenia may be closely tied to the way in which the lung carries out its function. It is not by accident that the cascading branches of the bronchial tree become smaller and smaller, nor is it good fortune alone that ties the dynamics of our every breath to this biological structure.

I shall argue that the lung is made up of fractal processes not unlike the heart, some dynamic while others are now static. However, both kinds of processes lack a characteristic scale and I give a simple argument establishing that such lack of scale has evolutionary advantages. In other words, a certain kind of scale-free complexity has helped us climb to the top of the evolutionary ladder, and understanding this kind of complexity may help us proceed to the next rung.

Relationships that depend on scale have profound implications for physiology. Consider the fact that the mass of a body increases proportionately with volume, the cube of the linear dimension, but the surface area increases as the square of the linear dimension. According to this principle of geometry, if one species is twice as tall as another, it is likely to be eight times heavier, but to have only four times as much of surface area. This tells us that the larger plants and animals must compensate for their bulk; respiration depends on the surface area for the exchange of gases, as does cooling by evaporation from the skin, and nutrition by absorption through membranes. One way to add surface area to a given volume is to make the exterior more irregular, as with branches and leaves on the trees, and in the bronchial tree in the mammalian lung, this is done with 300 million air sacs.

On a structural level, the notion of self-similarity can also be applied to other complex physiological networks. The vascular system, the blood vessels running through the human body, like the bronchial tree, form a ramifying network of tubes with multiple scale sizes. Many other fractal-like structures in physiology are also readily identified by their multiple levels of the self-similar branching and folding, for example, the bile duct system, the urinary collecting tubes in the kidney, the convoluted surface of the brain, the lining of the bowel, neural networks, and the placenta.[51]

In this section, my discussion is restricted to the time-independent structure of the lung, a discussion relating inverse power-law scaling in the size of successive generations of the bronchial tree, along with fractal geometry to anatomy. Such anatomical structures are only static in that they are "fossil remnants" of a morphogenetic process. As the lungs grow in the fetus, the separate branches first appear as buds, and then they develop and branch out, leaving the bronchial tree to be observed as the end product of the dynamical growth process. It is reasonable to suspect that morphogenesis itself could also be described as a fractal process, but one that is time-dependent. The question then arises as to how the structure of the lung is encoded within cells. It is possible that the information regarding the structure of the lung is encoded in the same manner that the pixels of a photograph are encoded within a computer. The photograph can then be reconstructed by reading the sequence of zeros and ones using a decoding program. However, encoding an information compression algorithm rather than the information itself would be much more efficient. A fractal growth process would imply a minimal code for the development of complex but stable structures. The code would be minimal because it would only require coding the mapping leading to the structure and not coding the structure itself. It is the difference between recording all the natural numbers or recording the rule by which the numbers can be generated. The itemized list would be infinitely long, but the rule merely states that we add one to each successive integer, starting from zero.

The success of the fractal models of the lung, the heart and other anatomical structures suggests that nature prefers fractal structures to those generated by classical scaling. Fractal structures are ubiquitous in physiology because they are more adaptive to internal changes and to changes in the environment, than are structures generated by other scaling algorithms. In fact, I constructed a simple quantitative model of error response to illustrate the difference between classical and fractal models. In Fig. 5.10, the growth of error associated with fluctuations in the parameter controlling the size of the diameter of the bronchial airway within two such models is shown. The error in the classical model, produced by chemical changes, environmental changes, or genetic mutations, grows geometrically. In this figure, small errors in each generation having a variance of 0.01 were

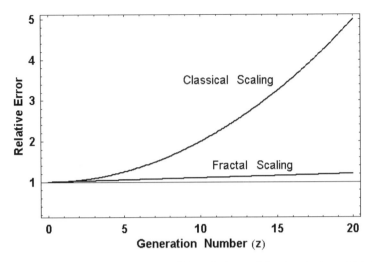

Fig. 5.10: The relative error between model predictions of the size of a bronchial airway and the same prediction when the growth is corrupted by a noisy parameter. The difference between the classical-scaling model and the fractal-scaling model[73] is striking.

introduced, and by the sixteenth generation of the bronchial tree, the predicted size of the airway, with and without error, differ by a factor of three. An organism with this sensitivity to error (or to fluctuations in the environment during morphogenesis) would not survive over many generations of the species.

On the other hand, the second curve in Fig. 5.10 shows that the fractal model is essentially unresponsive to error, starting with the same small errors in each generation. The fractal model is very tolerant of variability in the physiological environment. This error tolerance can be traced back to the fact that a fractal object has a distribution of scales, but is not dominated by any one scale. The scale-free distribution ascribes many scales to each generation in the bronchial tree, therefore any scale introduced by an error is already present, or nearly so, in the original tree. Thus, the fractal model is *preadapted* to certain physiological fluctuations such as genetic errors and variations in the chemistry of the growth environment.[73] This error tolerance insures the robustness of both fractal physiologic structures and processes, such that a fractal rule has evolutionary advantage over that of geometrical scaling.

As with the heart, the variability of breathing rate and breath-to-breath time intervals are both denoted by BRV, to maintain a consistent nomenclature. A breathing rate time series for a healthy subject was presented in Fig. 2.1 where the existence of a pattern in the BRV was observed by contrasting the original data with the same data randomly shuffled. Recall that shuffling the data destroys long-time correlations because such correlations depend on the time ordering of the intervals. However, this reordering of terms does not change the relative frequency of intervals, since intervals are neither added nor subtracted in the shuffling process, so the probability distribution remains the same. While it was apparent in Fig. 2.1 that a pattern existed in the original BRV data, it was not clear how one could quantify that pattern in order to use the information that the pattern contains.

An early application of fractal analysis to the breathing data was made by Szeto *et al.*[74] to fetal lamb breathing. The changing patterns of breathing in 17 fetal lambs and the clusters of faster breathing rates, interspersed with periods of relative quiescence, suggested to them that the breathing process was self-similar. The physiological property of self-similarity implies that the structure of the mathematical function describing the time series is repeated on progressively finer time scales. Clusters of faster rates were seen within the fetal breathing data, what Dawes *et al.*[75] named breathing episodes. When the time series were examined on even finer time scales, clusters could be found within these clusters, and the signature of this scaling behavior emerged as an inverse power law. Consequently, the fractal scaling was found to reside in the statistical properties of the fluctuations and not in the geometrical properties of the dynamic variable.

Examples of HRV and BRV time series data on which scaling calculations are based are shown in Fig. 5.11. A typical BRV time series for a senior citizen at rest is shown at the top of the figure; the simultaneous HRV time series for the same person is depicted

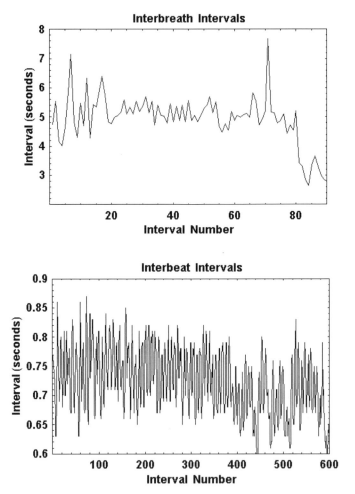

Fig. 5.11: Typical time series from one of the 18 subjects in the study conducted by West et al.,[76] while at rest, is shown for the interbreath intervals (BRV) and the interbeat intervals (HRV) time series.

at the bottom of the figure. As heart rate is higher than respiration rate, in the same measurement epoch, there is factor of five more data for the HRV time series than there is for the BRV time series. These data were collected under the supervision of my friend and colleague Dr. Richard Moon,[76] the Director of the Hyperbaric Laboratory at

Duke Medical Center. Looking at these two time series together, one is struck by how different they are. It is not apparent that both physiological phenomena scale in the same way.

The allometric aggregation method applied to the various time series obtained by West et al.[76] yields the typical results depicted in Fig. 5.12, where the logarithms of the aggregated variance versus the aggregated average are plotted for the HRV and BRV data depicted in Fig. 5.11. At the extreme left of each graph in Fig. 5.12 ($m = 1$), all the data points are used to calculate the variance and average, and at the extreme right, the aggregated quantities use $m = 10$ data points. Note that we stop the aggregation at ten points because of the small number of data in the breathing sequence. The solid curve at the top of Fig. 5.12 is the best least-square fit to the aggregated BRV data and has a slope of 0.86 which is the scaling index. A similar graph is constructed for the HRV data in the lower curve, where we obtain a slope of 0.80 for the scaling index.

The scaling index of both the HRV and BRV time series increase with increasing levels of exercise, but the data is not shown here, mainly because I can't show everything all the time. The 18 subjects in the experiment rode a stationary bicycle with various levels of load on the wheels to mimic cycling uphill. The breathing rate, breathing volume and heart rate were simultaneously monitored for each of the individuals in the study. The consistent result was that as the level of exercise increased, the amount of variability in both HRV and BRV decreased, indicating that the associated time series were becoming more ordered. This increase in scaling index was determined to be statistically significant.[76] The scaling indices and fractal dimensions obtained from these curves are consistent with the results obtained by other researchers.[77]

Such observations regarding the self-similar nature of breathing time series have been used in a medical setting to produce a revolutionary way of utilizing mechanical ventilators. Historically, ventilators have been used to facilitate breathing after an operation and have a built-in frequency of ventilation, as I discussed in an earlier chapter.

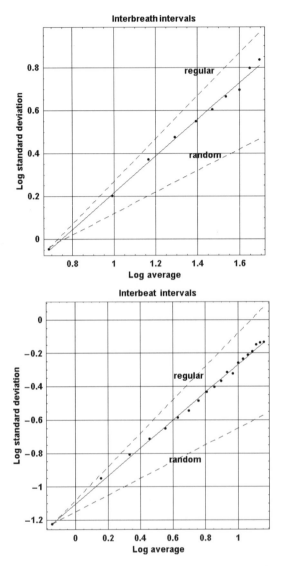

Fig. 5.12: A typical fit to the aggregated variance versus the aggregated mean for the BRV and HRV time series shown in Fig. 5.11. The points are calculated from the data and the solid curve is the best least-square fit to the data points. The upper curve is the best fit to the BRV data (*slope* = 0.86), the lower curve is the best fit to the HRV data (*slope* = 0.80). It is evident from these two graphs that the allometric relation does indeed fit both data sets extremely well and lies well within the regular (upper dashed line) and random (lower dashed line) boundaries.

This ventillator design has recently been challenged by Mutch et al.,[78] who used an inverse power-law spectrum of respiratory rate to drive a variable ventilator. They demonstrated that this way of supporting breathing produces an increase in arterial oxygenation over that produced by conventional control-mode ventilators. This comparison indicates that the fractal variability in breathing is not the result of happenstance, but is an important property of respiration. A reduction in variabilty of breathing reduces the overall efficiency of the respiratory system.

Altemeier et al.[79] measured the fractal characteristics of ventilation and determined that not only are local ventilation and perfusion highly correlated, but they scale as well. Finally, Peng et al.[80] analyzed the BRV time series for 40 healthy adults and found that under supine, resting, and spontaneous breathing conditions, the time series scale. This result implies that human BRV time series have long-range (fractal) correlations across multiple time scales.

Some particularly long records of breathing were made by Kantelhardt et al.[81] for 29 young adults participating in a sleep study where their breathing was recorded during REM sleep, non-REM sleep and periods of being awake. They showed that the stages of healthy sleep (deep, light and REM) have different autonomic regulation of breathing. During deep or light sleep, both of which are non-REM, the scaling index denoted as α was determined to be $\alpha = 1/2$, and consequently, the BRV time series was an uncorrelated random process. On the other hand, during REM sleep and periods of wakefulness, the scaling index was found to be in the interval of $0.9 > \alpha > 0.8$, such that the BRV time series has a long-time memory.

The evidence for scaling of BRV time series is fairly compelling, but until now, the source of that scaling, whether it is chaos or colored noise, has not been determined. In the previous section, I introduced ART (the attractor reconstruction technique) to assist in that determination regarding heart rate variability. I used the time series data plotted against itself to determine if any structure over and above what would occur due to random noise would reveal itself. In Fig. 5.13, the breath interval time series is plotted versus a copy of itself shifted by one breath. In this way, the two coordinates of a point in the plane are completely determined by the data. In the upper graph, the result of such a plot is recorded and in the lower graph, the same thing is done using randomly ordered (shuffled) data points.

Connecting the data points in Fig. 5.13 by straight lines facilitates seeing the time order in which the data points occur. The darkness of the pattern is an indication of more connecting lines crossing or lying near one another. The experimental BRV data, even with the relatively few data points obtained in the experiment, 691 breath intervals, shows a dynamical structure. Within the central region of the "attractor", there are regions where the trajectory tracing out the dynamics of the breathing pattern generator spends more time and is therefore darker than the surrounding area. It may not be obvious that the BRV data is generated by an attractor, but it is certainly the case that the data is distinctly different from uncorrelated random noise, which would result in a featureless ball in phase space. The internal structure, where the trajectory avoids certain central regions in the "attractor", is lost in the lower figure where the data points are shuffled. However, the shuffled data at this point do not give rise to a symmetric Gaussian blob of points.

Webber[82] also investigated the nonlinear physiology of breathing, pointing out that the central rhythm and pattern generators within the brainstem/spinal cord are subjected to afferent feedback and suprapontine feedforward inputs of varying coupling strengths. In short, there are multiple signals that influence one another in the respiratory controller and the dynamics are nonlinear, i.e. a given response is not proportional to a given stimulus. However, Webber applied a number of techniques from nonlinear dynamics systems theory to the analysis of breathing. He recorded 4500 consecutive respiratory cycles for the spontaneous breathing patterns of unanesthetized, unrestrained rats (10 in all). He then constructed

Fractal Physiology • 215

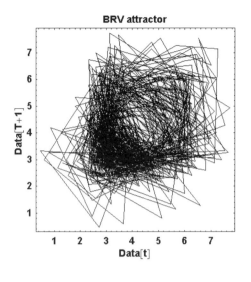

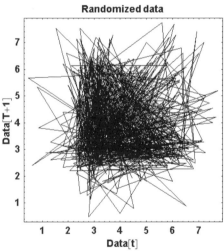

Fig. 5.13: The upper graph is the BRV data (691 breath intervals) plotted versus a copy of itself shifted in time by one breath. The lower graph is the same plot as the uppoer but with the ordering of the data point randomized.

phase-space plots that are similar but not identical to the one shown in Fig. 5.13, using continuous thoracic pressure fluctuations. His plots reveal patterns in the breathing data that are consistent with chaotic dynamics.

In addition, he used information (entropy) as a measure of the level of complexity of the respiratory signal and determined that the level of complexity increased with the level of activity of the rats.

5.6. Fractal gait

Walking is so natural a physical movement that we do not give it much thought. I only become conscious of my movements during gait when I slam my toe against the table in the dark and recall with some upset that I moved it during the day; when I trip over a footrest and remember an adolescent afternoon when my winning touchdown was discouraged by the school's star tackle; or the weight of years makes each step a painful experience. These considerations aside, in my prime, I walked confidently with a smooth pattern of strides and without open variation in my gait. This apparent lack of pattern is remarkable considering that the motion of walking is created by the loss of balance, as pointed out by Leonardo da Vinci (1452–1515) in his treatise on painting. da Vinci considered walking to be a sequence of fallings; hence, it should come as no surprise that there is variability in this sequence of falling intervals, even if such variability is usually masked.

The regular gait cycle, so apparent in everyday experience, is not as regular as scientists had believed. Gait is no more regular than is normal sinus rhythm or breathing. The subtle variability in the stride characteristics of normal locomotion was first discovered by the 19th century experimenter Vierordt,[83] but his findings were not followed up on for more than 120 years. The random variability he observed was so small that the biomechanical community has historically considered these fluctuations to be an uncorrelated random process. In practice, this means that the fluctuations in gait were thought to contain no information about the underlying motorcontrol process. The follow-up experiments to quantify the degree of irregularity in walking was finally done in the middle of the last decade by Hausdorff *et al*.[84] The latter experiments involved observations of healthy individuals, as well as of subjects having certain diseases that affect gait, and also the elderly, a discussion of pathology that I shall postpone until a later chapter. Additional experiments and analyses were subsequently done by West and Griffin[85–87] who both verified and extended the earlier results.

Human gait is a complex process since the locomotor system synthesizes inputs from the motor cortex, the basal ganglia and the cerebellum, as well as feedback from vestibular, visual and proprioceptive sources. The remarkable feature of this complex phenomenon is that although the stride pattern is stable in healthy individuals, the duration of the gait cycle is not fixed. Like normal sinus rhythm in the beating of the heart, where the interval between successive beats changes, the time interval for a gait cycle fluctuates in an erratic way from step to step. To date, the gait studies carried out concur that the fluctuations in the stride-interval time series exhibit long-time inverse power-law scaling, indicating that the phenomenon of walking is a self-similar fractal activity.

Walking consists of a sequence of steps and the corresponding time series is made up of the time intervals for these steps. These intervals may be partitioned into two phases: a stance phase and a swing phase. It has been estimated, using blood flow to skeletal muscles, that in the stance phase muscles consume three times as much energy as do the swing phase muscles, independently of speed.[88] The stance phase is initiated when a foot strikes the ground and ends when it is lifted. The swing phase is initiated when the foot is lifted and ends when it strikes the ground again. The time to complete each phase varies with the stepping speed. A stride interval is the length of time from the start of one stance phase to the start of the next stance phase. It is the variability in the time series made from these intervals that is being probed in this analysis of the stride interval time series.

One way to define the gait cycle or stride intervals is the time between successive heel strikes of the same foot.[86] An equivalent

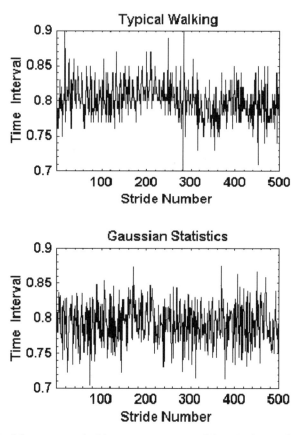

Fig. 5.14: Real data compared with computer-generated data. At the top the time interval between strides for the first 500 steps made by a typical walker in the experiment[85] is depicted. At the bottom a computer-generated time series having uncorrelated Gaussian statistics is shown, with the same mean and variance as in the data shown above.

definition of the stride interval uses successive maximum extensions of the knee of either leg.[85–87] The stride interval time series for a typical subject is shown in Fig. 5.14 where it is seen that the variation in time interval is on the order of 3 to 4%, indicating that the stride pattern is very stable. The stride interval time series is referred to as stride rate variability (SRV) for name consistency with the other two time series we have discussed. It is the stability of SRV that historically led investigators to conclude that it is safe to assume that the stride interval is constant and the fluctuations are merely biological noise. The second set of data in Fig. 5.14 is computer-generated Gaussian noise, having the same mean and standard deviation as the experimental data. Note that it is not easy to distinguish between the two data sets. However, the experimental data fluctuates around the mean gait interval, and although they are small, they are non-negligible because they indicate an underlying complex structure and these fluctuations cannot be treated an uncorrelated random noise.

Using an SRV time series of 15 minutes, from which the data depicted in Fig. 5.14 were taken, I apply the allometric aggregation procedure to determine the relation between the standard deviation and average of the time series as shown in Fig. 5.15. In the latter figure, the curve for the SRV data is contrasted with an uncorrelated

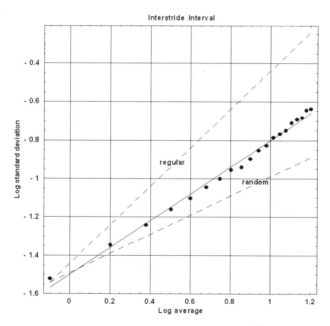

Fig. 5.15: The SRV data for a typical walker in the experiment[85] is used to construct the aggregated variance and mean as indicated by the dots. The logarithm of the variance is plotted versus the logarithm of the mean, starting with all the data points at the lower left to the aggregation of 20 data points at the upper right. The SRV data curve lies between the extremes of uncorrelated random noise (lower dashed curve), and regular deterministic process (upper dashed curve) with a fractal dimension of $D = 1.30$.

random process (*slope* = 0.5) and a regular deterministic process (*slope* = 1.0). The slope of the data curve is 0.70, midway between the two extremes of regularity and uncorrelated randomness. As in the cases of HRV and BRV time series, the erratic physiological time series are determined to be random fractal processes.

In a previous section, I argued that if the power-law index, the slope of the variance versus average curve on a log-log graph, is greater than one, then the data are clustered. In the SRV context that is indicated by a slope greater than the random dashed line, this clustering means that the intervals between strides, change in clusters and not in a uniform manner over time. This result suggests that the walker does not smoothly adjust his/her stride from step to step. Rather, there are a number of steps over which adjustments are made, followed by a number of steps over which the changes in stride are completely random. The number of steps in the adjustment process and the number of steps between adjustment periods are not independent. The results of a substantial number of stride interval experiments support the universality of this interpretation.

In today's world of science, we are no longer limited by the data we can collect on our own, in our own laboratory, or in the laboratory of a friend, since we can download data sets from other researchers working worldwide, who make their data available. For example, researchers at Beth Israel Hospital and Harvard Medical School, including J.M. Hausdorff, C.-K. Peng and A.L. Goldberger, post their data on the internet at http://www.physionet.org. This is a National Institutes of Health sponsored website, *Research Resource for Complex Physiologic Signals*. A number of these posted data sets are analyzed in *Biodynamics*[89] where additional support for the idea that the statistics of SRV time series are fractal can be found as well as on PhysioNet.

The SRV time series for 16 healthy adults were downloaded from *PhysioNet* and the allometric aggregation procedure carried out. Each of the curves looked more or less like that in Fig. 5.15, with the experimental curve being closer to the indicated regular or the random limits (dashed curves). On average, the 16 individuals have fractal dimensions for gait in the interval of $1.2 \leq D \leq 1.3$ and an average correlation on the order of 40%.[90] The fractal dimension obtained from the analysis of an entirely different dataset, obtained using a completely different protocol, yields consistent results. The narrowness of the interval around the fractal dimension suggests that this quantity may be a good quantitative measure of an individual's dynamical variability. I suggest the use of the fractal dimension as a quantitative measure of how well the motorcontrol system is doing in regulating locomotion. Furthermore, excursions outside the narrow interval of fractal dimension values for apparently healthy individuals may be indicative of hidden pathologies.

Further analysis was done on the SRV time series of fifty children, also downloaded from *PhysioNet*. One of the features of the time series were large excursions in the fractal dimension for children under the age of five, $1.12 \leq D \leq 1.36$, unlike the narrower range of values for mature adults. The interval expands from 0.08 for adults to 0.24 for children, a factor of three decrease of the interval from childhood to adulthood. It is clear that the average fractal dimension over each group is the same, approximately 1.24, but the range of variation in the fractal dimension decreases significantly with age.[89,90] This would seem to make the fractal dimension an increasingly reliable indicator of the health of the motorcontrol system with advancing age.

I have so far shown results for three types of neurophysiologic control systems, two involuntary (heartbeat and breathing regulation) and the remaining one which is voluntary (gait regulation). The time series for human heart rate, breathing rate and stride rate have each had "random" fluctuations that scale in time and are therefore described as fractal statistical phenomena. The first two have also revealed properties associated with chaotic time series, implying that nonlinear dynamical equations may be the source of the fractal fluctuations. The ART technique is here applied to the SRV time series as shown in Fig. 5.16. In the graph, it is clear that the SRV data lie on

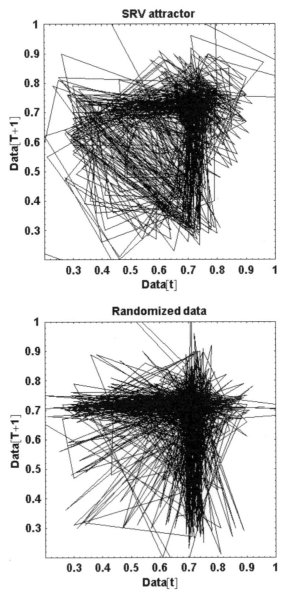

Fig. 5.16: The upper graph is the SRV data (1252 step-to-step intervals) plotted versus a copy of itself shifted in time by one step. The lower graph is the same plot but with the ordering of the data point randomized.

an "attractor" with two major branches that are perpendicular to one another, with a third branch joining them to form a triangle in phase space. The darkness of the branches indicate where the dynamics of the stride interval regulator spends most of its time. This attractor is in stark contrast to random data points with Gaussian statistics, that would concentrate in a featureless ball centered on the mean and with a diameter determined by the variance of the fluctuations. Shuffling the data leads to the reconstruction in the lower figure, where the hollow in the center of the "attractor" is lost, but a great deal of structure persists.

It should not go unnoticed that people basically use the same control system when they are standing still, maintaining balance, as when they are walking. This observation would lead one to suspect that the body's slight movements around the center of mass of the body, that ficticious point at which all of the body's mass is located in any simple model of locomotion, would have the same statistical behavior as that observed during walking. These tiny movements are called postural sway in the literature and have given rise to papers with such interesting titles as "Random walking during quiet standing" by Collins and De Lucca.[91] It has been determined that postural sway is really chaotic,[91,92] so one might expect that there exists a relatively simple dynamical model for balance regulation that can be used in medical diagnosis. Again, the fractal dynamics can be determined from the scaling properties of postural sway time series, and it is determined that a decrease of postural stability is accompanied by an increase of fractal dimension. The application of this idea is described more completely in a later chapter.[92]

5.7. Fractal temperature

Since the beginning of the murder mystery genre, the time at which the victim dies has been one of the key factors in the story. The time of death determines the time at which the usual suspects, once they have been identified, need to account for their whereabouts. A clever deception concerning the time of death on the part of the perpetrator often sends the detectives off in the wrong track and thereby hangs the tale. Yet, what is the scientific basis for the determination of the time of death?

The notion that one can medically determine the time of death is based on a number of simple observations. First of all, the body temperature of every healthy person falls within a narrow range (96.7 to 99°F orally and half a degree higher rectally) and a change in the body's core temperature by a few degrees is often fatal. Secondly, the rate at which the body temperature is lowered, due to the loss of residual body heat, is constant over many hours, depending only on the difference in temperature between the body and the immediate environment. This rate of cooling is determined by Newton's law of cooling, which is based on observations of how inanimate objects lose heat over time, which is what the human body becomes once the spark of life has been extinguished. Finally, an accurate determination of the present core temperature of a human body, e.g. the temperature of the liver, enables the medical examiner to extrapolate backwards to determine how long it has been since the light went out.

The above way of determining the time of death is simply plain vanilla. While alive, the human body loses heat by radiation, convection, conduction and evaporation. After death, I rule out conduction as a mechanism for body cooling, unless the body is lying in the snow, and I eliminate evaporation unless the corpse and clothing are wet. This leaves radiation and convection. It has been determined empirically that the change in body temperature with time has a sigmoidal shape because of the body's extended surface. This differs from Newton's law of cooling in which the rate of heat loss is proportional to the simple difference between the body and the ambient temperatures, and is independent of geometry. A sigmoid curve for body cooling is depicted in Fig. 5.17, where the initial difference between body and ambient temperatures stays nearly constant at a

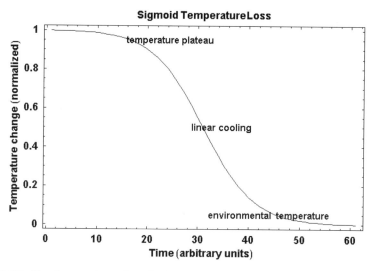

Fig. 5.17: The three regions of an idealized sigmoid curve of body cooling are indicated. The curve denotes the drop in the difference between normal body temperature and ambient temperature indicated as one, down to the environmental temperature indicated as zero.

plateau for a period of time, typically a few hours. This plateau is followed by an interval of essentially linear cooling, whose rate, the slope of the sigmoid curve, subsequently brakes and approaches zero when the temperature of the body approaches the temperature of the environment. The reasons for these various stages do not concern us, except in so far as to indicate that temperature regulation, even after death, appears to be a fairly well-understood phenomenon. Thus, the Medical Examiner can supply the information needed by the police to narrow the list of suspects. The medical assumption is that the average body temperature is one of the most comprehensive measure of the health-state of the body, both before and after death.

I now turn from the murder mystery back to the fate of the living. Due to my shift in purpose, I examine the first of the above assumptions; that body temperature is so well regulated that it is essentially constant in time and is therefore adequately represented by its average value. It is well established that throughout the day, the "normal" body temperature changes, being lowest in the early morning (4–6 a.m.), when it can be as low as 97°F and highest in the early evening (6–8 p.m.), when it can be as high as 99°F and still be considered normal. This is a circadian rhythm (periodic daily change) of temperature that is superposed on the normal erratic fluctuations observed in the body temperature.

Consequently, a test as to whether the hypothesis concerning physiological phenomena pertains to human body temperature time series in the same way as it did to HRV, BRV and SRV time series is needed. One limitation of such a study is the relative difficulty in obtaining long time series for body temperature. Fortunately, I was able to find a short time series that Buzz Hoagland recorded in June of 1998 and put on the Internet for his biology class. He recorded his own body temperature at 15-minute intervals for a few days whilst he was sick with bronchitis at home. Figure 5.18 displayed his body temperature time series for June 28 of that year, just when he was recovering from his illness and recorded his body temperature variability (BTV).

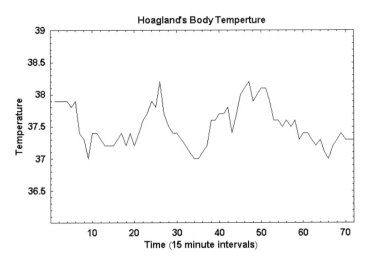

Fig. 5.18: The body temperature of Dr. Buzz Hoagland at 15-minute intervals for June 28, 1998. http://biology.wsc.ma.edu/biology/experiments/temperature/fever/. The last day of his bout with bronchitis.

Dr. Hoagland interpreted the variability in his body temperature time series as being due to the immune system waging war with the bronchial infection, an interpretation of variability based on competition between two mechanisms. On the other hand, I associate this erratic behavior of BTV with the natural variability of the body's thermal regulation. In addition to the erratic variation in the body temperature, there is a periodic oscillation that may be associated with circadian rhythm. To verify the conjecture regarding the temperature control system, that the erratic fluctuations in body temperature are associated with the healthy variability of the underlying thermoregulatory system, I apply the allometric aggregation approach to these 72 data points. Using the AAA method, the pattern shown in Fig. 5.19 is obtained. It is quite evident that the data points, plotted as the logarithms of the standard deviation and the average, with increasing aggregation number, lie along a straight line with a well-defined slope. The slope of the curve is a single parameter measure of the variable time series shown in Fig. 5.18. The slope

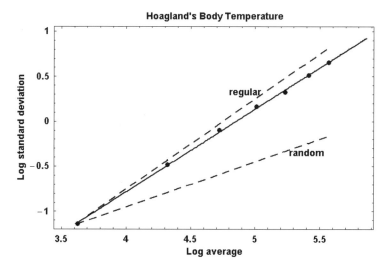

Fig. 5.19: The points depict the variance and average value obtained from the body temperature data depicted in Fig. 5.18 for a given level of aggregation, from no aggregation on the extreme left, to the aggregation of seven data points on the extreme right. The solid curve is the best fit to the data points and has a slope of 0.92.

of the curve is approximately 0.92, not too different from the value of 1.0 that occurs for regular time series, but sufficiently different to suggest an underlying fractal pattern. The nearness to a deterministic value may be the result of the circadian rhythm producing an additional non-stationarity in the time series, but in any event, it is clear that the fluctuations are not produced by an uncorrelated random process.

The shortness of this data set may call into question the conclusions drawn about body temperature time series. Furthermore, it might be argued that I should have obtained a time series from a healthy individual to make the conclusion I have drawn unassailable. However, this book is not intended to be a scientific document, but more a sketch of ideas underpinning a new way of understanding medicine. Thus, it would be good if these results can be improved upon or shown to be invalid. In either case, I will have accomplished my purpose, which is the re-examination of the ideas upon which medical understanding is built.

5.8. Fractal gut

I have noticed on days when I miss lunch that just before dinner, my stomach gives off remarkable sounds that can be quite embarrassing. Like the sounds of the heart and breathing, stomach rumbling is characteristic of its function. The main function of the stomach is associated with periodic contractions that help to mix, grind, and move the food toward the small bowel. These contractions are intimately related to spontaneous, rapid intracellular electrical depolarization caused by the exchange of ions through the membrane of smooth muscle cells. This signal is a precursor to stomach wall muscle contractions and while the mechanism of propagation of gastric electrical activity from one cell to another is still unknown, it is believed that periodic waves of depolarization and repolarization originate from a pacemaker area located in the corpus around the greater curvature of the stomach and spreads aborally toward the pylorus. The contractions are initiated by the vagus nerve. The visceromotor part of the vagus nerve innervates ganglion neurons that go to the lung for bronchoconstriction, to the heart for slowing the heart rate and to the stomach for secretion and constriction of smooth muscle.

The question arises whether these sounds and/or the associated electrical activity contain any useful information about digestion and whether cutaneous recordings of gastric electrical activity called electrogastrography (EGG) contain patterns that can be used to characterize physiological gastric activity, as well as to diagnose gastrointestinal pathologies. The first EGG measurement was made by Alvarez in 1921, and during the last several years, this technique has received renewed interest. To a large extent, this resurgence of interest can be attributed to the noninvasive character of this experimental method. While there have been several attempts to apply EGG for clinical assessment, most approaches involved qualitative and not quantitative evaluation of the recorded tracings. Recently, we hypothesized that the time series generated by gastric activity is a statistical fractal, and consequently ought to have a long-time memory that can be measured by means of a scaling parameter. My colleague and friend Mirek Latka set out to establish the scaling behavior in the variability of EGG tracings depicted in Fig. 5.20 thereby, quantifying its fractal character.

Using some techniques known as wavelet analysis we were able to determine a time series of the peak-to-peak intervals in gastric activity, which for notational consistency with other types of physiologic variability, we refer to as gastric rate variability (GRV).[93] The GRV time series is denoted by the curve in Fig. 5.21, where the separate intervals are joined by straight lines to aid the eye in determining the variability of the underlying process.

My colleagues and I determined that the EGG, as characterized by GRV time series, scales in time, and is a random fractal with an

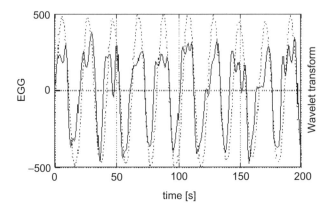

Fig. 5.20: First 200 seconds of an EGG recording of a healthy subject (solid line). The displayed waveform is the average of signals acquired form three bipolar electrodes. The wavelet transform of the EGG waveform, the GRV time series is given by the dashed line.[93]

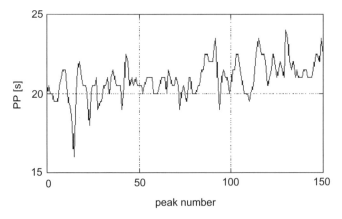

Fig. 5.21: Time series of the peak-to-peak intervals of the EGG waveform in Fig. 5.20. This is the GRV time series.[93]

average scaling index of $\alpha = 0.80 \pm 0.14$, corresponding to a fractal dimension of $D = 1.20 \pm 0.14$. In terms of random walk models, the phenomenon of stomach contractions is persistent, i.e. a random walker is more likely to continue in the direction of the preceding step than in changing directions. In the present context, this implies that if the time interval between the last two contractions was long, the subsequent interval will favor longer over shorter contractions.

This means that the GRV time series is not completely random, nor is it the result of processes with short-term correlations. Instead, the power-law form of the correlation function reveals that stomach contractions at any given time are influenced by fluctuations that occurred tens of contractions earlier. This behavior is a consequence of the fractal nature of GRV time series, the data showing long-time correlations that were statistically significant above the $p = 0.01$ level for each of the 17 subjects that contributed their variability to the study.[93]

5.9. Fractal neurons

Up to this point, the discussion has been focused primarily on time series generated by various complex physiological phenomena. Now let me examine a class of basic building blocks used to construct these phenomena, the individual neurons in the various control systems of the body. The neuron is in most respects quite similar to other cells in that it contains a nucleus and cytoplasm. However, it is distinctive in that long, threadlike tendrils emerge from the cell body, and those numerous projections branch out into still finer extensions. These are the dendrites that form a branching tree of ever more slender threads not unlike the fractal trees discussed earlier. One such thread does not branch and often extends for several meters, even though it is still part of a single cell. This is the axon that is the nerve fiber in the typical nerve. Excitations in the dendrites always travel essentially toward the cell body in a living system, whereas in the axon, excitations always travel away from the cell body.

The activity of a nerve is invariably accompanied by electrical phenomena. The first systematic observation of this effect was made by Luigi Galvani in 1791. He observed that when a muscle (frog's leg) was touched with a scalpel and a spark was drawn nearby, but not in direct physical contact with the scalpel, the muscle contracted.

In 1852, the German physician/physicist Helmholtz first measured the speed of the nerve impulse by stimulating a nerve at different points along the muscle and recording the time it took the muscle, to which it was connected, to respond. Electrical impulses are observed along the corresponding axon, whether it is an external excitation of a nerve or the transmission of a message from the brain. The properties of individual nerves seems to be ubiquitous because there is apparently no fundamental difference in structure, chemistry or function between the neurons and their interconnections in man and those in a squid, a snail or a leach. However, neurons do vary in size, position, shape, pigmentation, firing patterns and chemical substances by which they transmit information to other cells. The differences in neuron firing patterns are depicted in Fig. 5.22.

The rich dynamic structure of the neuron firing patterns has lead to their being modeled by nonlinear dynamical systems. In the physical neurons, these different patterns result from differences in the types of ionic currents generated by the membrane of the cell body of the neuron. At the top of Fig. 5.22 is a R2 neuron that is

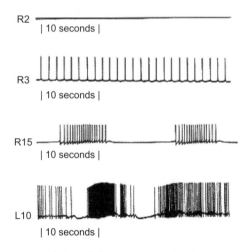

Fig. 5.22: Firing patterns of identified neurons in *Aplysia's* ganglion are portrayed. R2 is normally silent, R3 has a regular beating pattern, R15 a regular bursting rhythm and L10 an irregular bursting rhythm. L10 is a command cell that controls other cells in the system. [From Kandel,[94] Fig. 4, with the permission of *Scientific American*].

typically silent, i.e. the neuron does not fire on its own. If I think about this neuron in terms of dynamical behavior, I would classify it as a fixed point. The attractor structure is a fixed point, meaning that once the system has achieved this point in phase space, it does not move away from it on its own. The R3 neuron has a regular beating rhythm with a constant interval from one spike to the next. This pattern could be generated by a limit cycle in phase space, with the neuron firing at the completion of each cycle. The R15 neuron shown has a regular bursting rhythm that is a burst of activity, followed by a quiescent period, and another burst of activity. Within a given burst, the spikes are not equally spaced; however, they bunch up in the middle and spread out at the ends of the burst. The final neuron shown L10 has an irregular bursting rhythm, with unequal times between bursts and unequal intervals between spikes within a burst. The R15 and L10 neuron time series are not unlike the time series generated by certain chaotic attractors.

There are two different ways in which neurons can be fractal. The first way is through their geometrical structure. The shape of nerve cells may be fractal in space [e.g. see Fig. 3.3(b)], just as we observed for the cardiac conduction system and the architecture of the lung. The fractal dimension has been used to classify the different shapes of neurons and to suggest mechanisms of growth responsible for these shapes. The second way neurons can be fractal is through the time intervals between the action potentials recorded from nerves, as was observed for the interbeat interval distribution in cardiac time series and the interbreath interval distribution for breathing. The statistical properties of these time intervals have the same strange properties observed earlier, in that collecting more data does not improve the accuracy of the statistical distribution of the intervals measured for some neurons as characterized by the width of the distribution. The realization that the statistics of these intervals are fractal helps in understanding these surprising properties. Bassingthwaighte *et al.*[47] reviewed how the fractal dimension can be used to classify neurons into different functional types.

The statistical properties of the time intervals between action potentials display three different types of distributions: (1) Some neurons are well described by a *Poisson process*, in which the probability that there is an action potential in a time interval Δt is proportional to the size of the time interval. The durations of subsequent intervals are statistically independent. (2) Some neurons fire almost, but not exactly, at a constant rhythm. These are well described by a *Gaussian model*, where the intervals have a small dispersion around the mean value. (3) Some neurons show self-similar fractal patterns on the *scaled interval distributions*. In the last case, neurons occasionally have very long time intervals between action potentials. These long intervals occur frequently enough that they can be the most important in the determination of the average interval length. As more and more data are collected, longer and longer intervals are found. Thus, the average interval increases with the duration of the data record analyzed, in such a way that if an infinitely long record could be analyzed, then the average interval would be infinite. Such infinite moments are characteristic of stable Lévy processes, as discussed earlier.

Gernstein and Mandelbrot[95] were the first to quantitatively analyze the scaling properties of neuronal action potentials and a number of conclusions could be drawn from their analysis.

The first conclusion concerns the fact that as more and more data are analyzed, the values found for the average interval and the variance in the average increase, i.e. in the limit of an infinitely long data record, both the average and variance can become infinite. When the variance is infinite, the averages using different segments of the same data record may be markedly different. It is commonly thought that if the moments such as the average, are constant in time, calculated using data then the parameters of the process that generated the data are constant. Furthermore, if these moments vary, it is believed that the parameters that generate the data also vary. This commonly held notion is invalid. Processes, especially fractal processes, where the generating mechanism remains unchanged, can yield time series whose moments such as the average vary with time.

The second conclusion concerns the fact that as additional data are analyzed, increasingly long time intervals between pulses are found. Hence, the inclusion of these time intervals increases rather than decreases the variance in the measured distribution. The statistical irregularities in the measured distribution become larger as more data are collected. As stated by Gernstein and Mandelbrot[95]:

> Thus, in contradiction to our intuitive feelings, increasing the length of available data for such processes does not reduce the irregularity and does not make the sample mean or sample variance converge.

Action potentials can be recorded from the primary auditory neurons that transmit information about sound and from the pulse vestibular neurons that transmit information about head position to the brain. Fractal behavior is ubiquitous in these and other such sensory systems.[96] Without including the references given by Teich et al.,[96] we quote their review of the evidence for the observation of the fractal behavior:

> Its presence has been observed in cat striate-cortex neural spike trains, and in the spike train of a locust visual interneuron, the descending contralateral movement detector. It is present in the auditory system of a number of species; primary auditory (VIII-nerve) nerve fiber in the cat, the chinchilla, and the chicken all exhibit fractal behavior. It is present at many biological levels, from the microscopic to the macroscopic; examples include ion-channel behavior, neurotransmitter exocytosis at the synapse, spike trains in rabbit somatosensory-cortex neurons, spike trains in mesencephalic reticular-formation neurons, and even the sequence of human heartbeats. In almost all cases the upper limit of the observed time over which fractal correlations exist is imposed by the duration of the recording.

It would probably be considered bad form in a discussion of neurons and action potentials, if we did not comment on the electrical properties of the human brain. It has been known for more than a century that the activity of a nerve is based on electrical phenomena and that the mammalian brain generates a small but measurable electrical signal. The electroencephalograms (EEG) of small animals were measured by Caton in 1875 and those of man by Berger in 1925. The mathematician N. Wiener, whom I mentioned earlier as the father of cybernetics, thought that *generalized harmonic analysis* would provide the mathematical tools necessary to penetrate the mysterious relations between EEG time series and the functioning of the brain. The progress along this path has been slow and the understanding and interpretation of EEG remains quite elusive. After 130 years, one can only determine intermittent correlations between the activity of the brain and that found in EEG records. There is

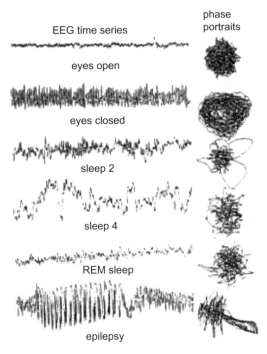

Fig. 5.23: Typical episodes of the electrical activity of the human brain as recorded from the electroencephalogram (EEG) time series together with the corresponding phase space portraits. The portraits are two-dimensional projections of the actual attractors. [From Babloyantz and Destexhe[97] with permission].

no taxonomy of EEG patterns that delineates the correspondence between those patterns and brain activity (see Fig. 5.23).

The traditional methods of analyzing EEG time series rely on the paradigm that all temporal variations consist of a superposition of harmonic and periodic vibrations, in the tradition of Wiener. The reconstruction technique ART introduced earlier, however, reinterprets the time series as a multidimensional geometrical object generated by a deterministic dynamical process in phase space. If the dynamics are reducible to deterministic laws, then the phase portraits of the system converge toward a finite region of phase space containing an attractor. In Fig. 5.23, EEG time series are depicted for a variety of brain states, including quiet resting with eyes open and closed, three stages of sleep and a petit mal epileptic seizure. Adjacent to each of these time series is depicted a projection of the EEG attractor onto a two-dimensional subspace using ART.[97]

The brain wave activity of an individual during various stages of sleep is depicted in Fig. 5.23. The standard division of sleeping into four stages is used. In stage one, the individual drifts in and out of sleep. In stage two, the slightest noise will arouse the sleeper, whereas in stage three, a loud noise is required. The final stage, level four, is one of deep sleep. This is the normal first sequence of stages one goes through during a sleep cycle. Afterwards, the cycle is reversed back through stages three and two, at which time dreams set in and the individual manifests rapid eye movement (REM). The dream state is followed by stage two, after which the initial sequence begins again. It is clear that whatever the form of the cognitive attractor, if such an object exists, it is not static, but varies with the level of sleep. In fact, the fractal dimension decreases as sleep deepens, from a fractal dimension of eight during REM to half of that in level four of deep sleep.[97] The dimension drops further to a value of approximately two during petit mal epileptic seizure. The seizure corresponds to highly organized discharges between the right and left hemispheres of the brain.

5.10. Internetwork interactions

Time series measurements of stride rate variability (SRV), breath rate variability (BRV), heart rate variability (HRV), gastric rate variability (GRV) and body temperature variability (BTV), all reveal a combination of randomness and order, characteristic of fractal random processes. The underlying control systems for each of these phenomena, consisting of neurons, also shares these fractal properties. I could look at these various time series in pairs or all together, to determine how the underlying physiologic dynamics may influence one another. However, such investigations are on the frontier of medicine and so I will leave such analyses to the specialists, but discuss instead on what has been determined from a more restricted set of comparisons.

The long-time memory revealed by the fractal scaling in the SRV, BRV, HRV, GRV, BTV and neuron time series, has been shown by a variety of nonlinear techniques to be the output of nonlinear dynamical systems. There is no precedent in classical physiology to explain such complex behavior, since this behavior contradicts the homeostatic notion that physiologic regulation should consist of damping such fluctuations and returning the system back to its equilibrium-like state. The dynamics introduced does provide feedback which is not of the traditional linear kind. It is a kind of fractal dynamics that arises due to the multi-scale nature of physiology. This fractal physiology is a universal characteristic of integrated neural control networks, where the regulating mechanisms interact as part of coupled cascades of feedback loops in systems far from equilibrium,[98] which is discussed more fully in the following chapters.

The respiratory and cardiac control systems are linked in their roles in maintaining homeostasis and responding to the body's circulatory needs. Many studies have documented the influence of breathing on heart rate. One kind of study concerned respiratory sinus arrhythmia in which the breathing cycle modulates the heart rate[99] and the heart rate variability changes due to changes in the breathing pattern.[100] In Calabrese's study,[100] the influence of resistive load such as exercise, in the forms of riding a bike or jogging, on the cardiorespiratory interaction was examined and found to increase the respiratory period and HRV. Another study examined slow breathing in different meditative states, breathing which induced extremely prominent heart rate oscillations in healthy young adults.[101] Yet, another study examined obstructive sleep apnea, a phenomenon where the periodic cessation of breathing during sleep, due to intermittent airway obstruction, alters healthy heart rate dynamics.[102] Each of these examples focused on the influence of slow regular breathing on heart rate; these experiments did not examine how the fluctuations in normal breathing are related to the fluctuations in normal heart rate.

An effect complementary to respiratory sinus arrhythmia is coupling between the cardiovascular and respiratory systems (cardioventrilatory system). This coupling was investigated by Gattlletly and Larsen,[103] who quantified the influence of HRV on respiratory rhythm. They examined patterns in the time series constructed from the time interval between the electrocardiograph QRST waves and onset of inspiration; patterns of correlation that appear during rest, sleep and under general anesthesia. Neither the above respiratory sinus arrhythmia studies nor the cardioventilatory studies examined the correlation between the fluctuations in heart rate and the fluctuations in breathing rate in awake, healthy individuals. Neither did they examine how such possible correlations might change under different respiratory loading conditions, such as different levels of exercise. One respiratory sinus arrhythmia study attempted to determine if the origin of fractal HRV was due to the fractal nature of respiration. Yamamoto et al.[104] examined the cross-correlation of the spectral indices for the inverse power-law spectra of HRV and BRV and concluded that the fractal behavior of the two processes were not correlated in any statistically significant way for subjects at rest. A more recent study by Peng et al.[80] quantified the fractal dynamics of human respiratory dynamics as a function of gender and healthy aging, but not on respiratory load.

The analysis of West et al.[76] showed, like the work of Yamamoto et al.,[104] that the fluctuations in heartbeat intervals and interbreath intervals are statistically independent of one another. Thus, even though there is often a slow modulating influence of one system on the other, the rapid scaling control within the separate systems remains independent, one from the other, or so it would appear.

Chapter 6

Complexity

Complexity has long been recognized as a barrier to understanding a large variety of naturally occurring phenomena, such things as the turbulent white water in mountain streams, the burning of fuel during a rocket launch, the mud that buries homes under slides in the hills of Los Angeles and the swarming of insects on a summer night. In a medical context, the complexity barrier is often encountered when attempting to understand physiological phenomena from a holistic perspective, rather than looking at specific mechanisms. I have used the allometric aggregation approach to establish that such dynamic phenomena are complex, at least in the sense that they generate time series that are statistical fractals. The scaling behavior of such time series determines the overall properties that such complex systems must have, similar to the older analysis of errors and noise in physical systems. The historical view of complexity involved having a large number of variables, each variable making its individual contribution to the operation of the system and each variable responding in direct proportion to the changes in the other variables. The small differences in the contributions produced the fluctuations in the observed outcome. The linear additive statistics of measurement error or biological noise is not generally applicable to complex medical phenomena. The elements in complex physiological systems are too tightly coupled, so instead of a linear additive process, nonlinear multiplicative statistics represent the fluctuations more accurately. In this chapter, I examine how intersystem interactions in a generic physiological system may give rise to the observed scaling.

The individual mechanisms giving rise to the observed statistical properties in physiological systems are very different, so I do not attempt to present a common source to explain the observed scaling in walking, breathing and the beating heart. On the other hand, the physiological time series for each of these phenomena scale in the same way, so that at a certain level of abstraction, the separate mechanisms cease to be important and only the relations matter but not those things being related. It is the relation between blood flow and heart function, locomotion and postural balance, breathing and respiration, that is important. The thesis of complexity theory, in so far as such a theory can be said to exist, is that such relations have a common form for complex phenomena. This assumption is not so dramatic as it might first appear. Consider that such relations have traditionally been assumed to be linear, in which case the control is assumed to be in direct proportion to the disturbance. Linear control theory has been the backbone of homeostasis, but it fails miserably in describing, for example, the full range of HRV from the running child to the sedate senior, with all the pathologies that await them along the way.

In this chapter, I present a general rationale for the variability seen in heartbeats, breathing, walking and the other physiologic systems that we have discussed. This rationale is not, however, intrinsically tied to medicine, except in so far as physiological phenomena can be said to be complex. It is the nature of complexity that is addressed here and not only complexity in physiology, even though it is possible to directly translate the arguments into a physiological context, which I occasionally do. I do not strictly follow the latter because such a presentation might obscure the generality of the argument and not properly credit the architects of the theory. The ideas for the approach presented here stem from systems theory, network theory and mathematics, but not necessarily and only incidently from medicine.

The approach I take is to review some of the basic properties of complex networks, beginning with the simplest, in which the connections among the nodes (the fundamental elements of the network) are random. Random networks are static and although they do demonstrate a number of interesting properties, they are too restrictive to mimic the full range of behaviors of most real-world phenomena. Random networks are characterized by average quantities, and in that regard, they share certain characteristics with medical doctrine. Two real-world properties that random networks do *not* capture are growth and the criteria for establishing new connections in growing networks. *Preferential attachment* is a rule by which newly formed nodes prefer to make connections to nodes having the greatest number of existing connections within a network. This rule, along with growth, is sufficient to develop an understanding of a class of scale-free networks. Such scale-free networks have inverse power-law distributions in the number of connections to a given node. There is a deep relationship between these scale-free networks and the fractal statistics discussed previously and I attempt to explain their relationship. For example, the notion of preferential attachment is a modern, albeit independent, rendition of Taylor's explanation of allometric scaling.

The issue finally addressed in this chapter is how to control complexity. In the present discussion, such control is identified as one of the goals of medicine, in particular, understanding and controlling physiological networks in order to insure their proper operation. The strategy I adopt is to distinguish between homeostatic control and allometric control mechanisms. Homeostatic control is familiar from earlier discussions having as its basis a negative feedback character which is both local and instantaneous. Allometric control, on the other hand, is a relatively new concept that can account for long-term memory, correlations that are inverse power law in time, as well as long-range interactions in complex phenomena as manifested by inverse power-law distributions in the system variable. Allometric control introduces the fractal character into the otherwise featureless random time series to enhance the robustness of physiological networks, by introducing either fractional Brownian motion or fractional Lévy motion into the control of the network.

It is not merely a new kind of control that is suggested by the scaling of physiologic time series. Scaling also suggests that the historical notion of disease, which has loss of regularity at its core, is inadequate for the treatment of the dynamical diseases encountered by today's physician. Instead of the loss of regularity, the loss of variability is identified with disease, so that a disease not only changes an average measure, such as heart rate, which it does in late stages, but is manifested in changes in heart rate variability at very early stages too. Loss of variability implies a loss of physiologic control and this loss of control is reflected in the change of fractal dimension, i.e. in the scaling index of the corresponding time series.

6.1. Random networks

In my discussion of the architecture of the human lung, I considered the fact that the bronchial airways bifurcate downward from the trachea for 23 generations, out to the alveoli. The diameter of the bronchial tubes decreases as an inverse power law in generation number and therefore lead to a fractal model for the scaling of the airways. In a similar way, the ramified structure of the cardiac conduction system was also found to be fractal. In fact, one of Mandelbrot's[57] ongoing contributions to science was his recognition of the existence of a substantial number of naturally occurring fractals, from neuronal trees whose structure is described by Rall's Law,[105] to botanical trees described by Murray's Law,[106] beginning five centuries ago with the observations of Leonardo da Vinci[107] (1452–1519) that successive branchings in a tree are related to one another by a simple mathematical relationship, a scaling relation.

MacDonald distinguishes between trees and networks.[108] Consider the branch of a system to be represented by a line and the intersection of two such lines to represent an interaction. Using this rule, a tree is an open structure with no closed loops and a network closes back on itself. Consequently, the control of a tree only proceeds in one direction, from the earlier to the later branches, whereas a network can have feedback by means of which later branches modulate the influence of earlier branches, influencing the overall function of the system. Of course, this distinction between a tree and a network is more apparent than real, as in the, bronchial tree. While it is true that the physical structure of the bronchial airways spread out into a fractal tree, the alveoli at the lung surface do connect with fractal capillary beds to exchange gases with the blood. If a lack of oxygen or an excess of carbon dioxide is detected in the blood, this information is fed via the sympathetic nervous system to the heart and lung, and stimulates the heart to beat faster and the lungs to work harder. Consequently, there is an effective closed loop for the lungs through their coupling to various other physiological systems or trees by means of chemistry and the nervous system. Thus, it is the nonlinear behavior of networks that is important in this chapter, systems that close back on themselves in some way for the purpose of regulation.

The present goal is to understand the properties of abstract networks independently of any particular realization, at least for the time being. Assume that the elements of a network are given by a number of dots placed on a page as shown in Fig. 6.1(a). Such dots might represent power generation stations spread across the country, or they might be airports in various cities or the neurons in a motor-control system. For our purpose, it does not matter what the nodes or elements of the network are, so we place them in a circle explicitly disregarding any geographical or geometrical effects of the system. To make this a network, we need to connect the dots, where the connections between the nodes are indicated by straight lines representing

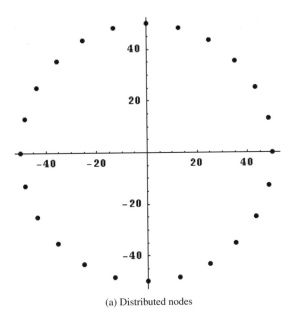

(a) Distributed nodes

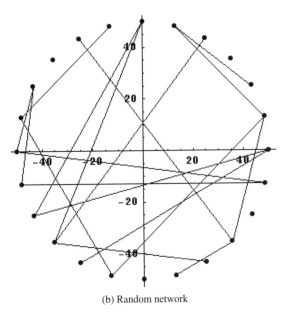

(b) Random network

Fig. 6.1: (a) Twenty-five points are placed uniformly on a circle to represent a distribution of nodes. (b) Twenty randomly selected pairs of nodes are connected to form a random network.

the nodal interactions. In this notation, the crossing lines do not represent interactions unless they pass through a node. Nodes that are not connected by a line have no direct influence on one another. Connecting airports by allocating restricted plane routes determines one of the country's major transportation grids. One could also examine the number of connections made with any particular node on the World Wide Web (WWW). Given the difference between the functions of the WWW and the air transportation grid, applying the old adage "form follows function", one would assume that the two have very different architectures. However, this turns out not to be the case. The connections that are made between the nodes is all-important, since the connections determine the mutual interactions among the nodes and these connections determine what the network does and not the properties of the individual nodes. Different choices of connections can have devastating consequences regarding how successfully the network realizes its function. Of course I am interested in the role played by the architecture of the network on the system's function in a medical context.

The simplest complex network can be constructed by ignoring any system-specific design strategy that would distinguish between a computer network, a transportation system or a homeostatic feedback loop within the body. One way to realize this goal is to devise a strategy to randomize the connections made within the network. For example, number the 25 nodes in Fig. 6.1(a) on pieces of paper, put the slips of paper into a bag, and randomly draw two from the bag and toss a coin. If the coin comes up heads, draw a line between the two numbered nodes to indicate that they interact with one another. If the coin comes up tails, do not draw a line and the two nodes are presumed to be independent of each another. Put the two slips of paper back into the bag and repeat the process. In this way, a network consisting of randomly connected nodes can be constructed. Figure 6.1(b) depicts such a simple random network consisting of 25 nodes at an intermediate stage of development, having only 20 connections out of a possible 300. It is evident from the sketches in Fig. 6.1 that it is very difficult to reach any general conclusions regarding networks with random connections by mere inspection. Can something of value to medicine be learned from such random networks?

In the mid 1950s and early 1960s, the Hungarian mathematicians Paul Erdös and Alfréd Rényi investigated the properties of random networks and were able to draw a number of remarkable conclusions. They applied Graph Theory to the study of random networks, a branch of mathematics dating back 300 years to Leonard Euler. Excellent accounts of the history of the Erdös-Rényi work on random graphs and the subsequent development of a theory of network connectivity is presented in a number of recent excellent popular books.[109,110,111]

Figure 6.1(a) depicts twenty-five independent nodes (elements) and Fig. 6.1(b) indicates a random linking of twenty pairs of these nodes. As connections are added to the network, it is observed that local clusters are randomly formed and these clusters are typically detached from one another, so that the network as a whole is made up of a number of these detached clusters. This clustering is more apparent when the network begins with a large number of nodes. However, as ever-more connections are added to the random network, a dramatic change in the network character eventually occurs, similar to the transformation that occurs in water during freezing. In water, the transition involves the short-range interactions of the particles in the fluid-phase changing their character to the long-range interactions of the particles in the solid-phase. Barabási[110] pointed out that in this latter stage of the random network, it is possible to reach any node from any other node by simply following the links. Here, all elements interacts with one another. Thus, there is a critical number of links in a random network at which point the average number of links, per node is one. Below the critical number of links, the network consists of disjoint clusters of interacting nodes. Above the critical number of connections, the random network is a single immense cluster in which all nodes participate.

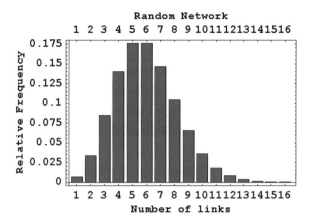

Fig. 6.2: The Poisson distribution for the number of links to a node in a random network is shown. The average number of links for the distribution shown is approximately five.

There are a number of other properties that were established for a random network. If the network is sufficiently large, then all nodes have approximately the same number of links. However, there are certainly deviations from the average. As shown in Fig. 6.2 where the relative number of connections per node is depicted, some nodes have more links than the average and some have fewer links than the average. The peak of the distribution, like the bell-shaped curve of Gauss, is at the average number of links. The distribution in the number of links to a given node in the random network of Erdös and Rényi is named after the French mathematician, Poisson.

Recall our earlier discussion of the law of errors and how the properties of errors stimulated 19th century scientists and philosophers to interpret social phenomena in terms of average values. The same interpretation again emerged in the 20th century from random network theory[110]:

> Erdös and Rényi's random universe is dominated by averages. It predicts that most people have roughly the same number of acquaintances; most neurons connect roughly to the same number of other neurons; most companies trade with roughly the same number of other companies; most Websites are visited by roughly

the same number of visitors. As nature blindly throws the links around, in the long run no node is favored or singled out.

The point being made here is reminiscent of the philosophy leading to the adoption of average value measures in medicine. Barabási[110] and Watts[111] argue that this view on the importance of average values applied to real-world networks, such as the World Wide Web and the Internet, has been shown to be wrong. The inapplicability of average values to these complex phenomena arise for the same reasons that they do not apply to medicine.

Complex phenomena are often overwhelming, presenting, as they do, a variety of variables interacting across multiple scales in space and time. Investigators try and make sense out of the many variables and interactions, but are often left in a quandary, confused about the cause and effect. The cyclic behavior of many complex phenomena is often obscured by strong fluctuations that appear to disconnect the present from the past. It is easy to understand how scientists choose to disregard faint patterns in the data and to assume that complex phenomena are random. After all, this approach to complexity has been taken by physicists for the past few hundred years as foundational in such disciplines as thermodynamics. It would not have been so remarkable, if not for the fact that this approach to complexity has been so successful as to seduce scientists into using randomness to "explain" non-physical phenomena.

However, once randomness is accepted as the proper description of complex phenomena, the subtle effects of complexity are suppressed by averaging procedures, so that only the gross influences of the average measures remain. The question then arises as to how the network theory got beyond the bell-shaped distribution in the number connections?

Random networks provide a comprehensive but limited view of the world. One of the most significant limitiations is illustrated by a social phenomenon now known as six degrees of separation and is discussed in those recent books[109–111] I mentioned earlier. I first heard of the six degree of separation phenomenon from

Elliott Montroll when I was a newly minted graduate in physics at the University of Rochester. Elliott gave a lecture at his own Festschrift in which he showed how close he was (five degrees of separation) to Benjamin Franklin. The proceedings of this meeting contains a picture of him giving the lecture with a caption that reads[112]:

> Montroll proves that he is one cheek-pinch and four hand-shakes away from Benjamin Franklin through the following sequence: Franklin pinched the cheek of the boy, John Quincy Adams, who shook hands with Joseph Henry, during congressional session December 1837, who shook hands with Hubert Anson Newton (at the 1864 National Academy Meeting–Newton being a Yale astronomer and best friends of J. Willard Gibbs), who shook hands with Lynde Phelps Wheeler (student of J. Willard Gibbs, Yale 1893–1899), who shook hands with Montroll (being introduced by the more famous Gibbs student, Edwin Bidwell Wilson at the time Wheeler was writing his Gibbs biography, 1949).

Another example of social distance that Montroll discussed with his students and post docs had to do with the random network mathematician Erdös and what is termed the Erdös number. We were told by Montroll that in the late 1960s, there was a famous Hungarian mathematician who was homeless, without permanent academic position, who traveled the world visiting other mathematicians to generate, collaborate and solve interesting problems in mathematics. This romantic image of a vagabond mathematical genius, roaming the world in search of problems, has stayed with me, although, I must say, my emotional reaction to the image has changed from envy to sadness over the years. Erdös was uniquely prolific, so much so that a number had been invented to measure the intellectual distance between other mathematicians and him. A person who collaborated with Erdös has an Erdös number of one; a person who collaborated with an Erdös co-author has an Erdös number of two and so on. It was fascinating to me when nearly 40 years later, after Montroll and Erdös had both died, that I read in *Linked*,[110] a list connecting the book's author Barabási with Erdös, with an Erdös number of four.

Included in that list was George H. Weiss, a colleague and friend of mine, with whom I have published a paper. George has an Erdös number of two, thereby bestowing on me an Erdös number of three. It is curious how even such an abstract connection gives one a sense of continuity with the past.

Barabási points out that the existence of such a quantity as the Erdös number emphasizes how interconnected the scientific community actually is, which is another manifestation of the six degrees of separation phenomenon discovered in society at large. Within the physical sciences, this phenomenon has metamorphisized into the "small-world" theory, exemplified by the typically small values of the Erdös number. The small-world theory began as a theory of social interactions in which social ties can be separated into two primary kinds: strong and weak. The strong ties exist within a family and among closest friends, those that you call in the event of emergency and contact to tell when you secure a promotion. These are the people that make up most of our lives. There are also the weak ties such as most colleagues at work, friends of friends, business acquaintances and so on.

Clusters form for closely knit groups, just as in random networks, clusters in which everyone knows everyone else. These clusters are formed from stong ties, but the clusters are coupled to one another through weak social contacts. The weak ties provide contact from within a cluster to the outside world. It is the weak ties that are all important for interacting with the world at large, e.g. for getting a new job. A now classic paper by Granovetter, *The strength of weak ties*,[113] explains how it is that the weak ties to near strangers are much more important in getting a job than are the stronger ties to one's friends.

The next stage in the progession of network development is the synthesis of clustering and random networks, such as was done by Watts and Strogatz.[114] They approached the problem from the perspective of understanding the phenomenon of the synchronization of complex phenomena. Examples include the spontaneous organization of clapping into rhythmic pulses at a concert and the orchestration of the chirping of crickets. The model of Watts and Strogatz is rather straightforward as we view it now, but it was revolutionary at the time when it was introduced. First, consider a simple model of a social network in which the links are given by the nearest neighbor and the next nearest neighbor interactions as shown by the schematic drawing on the left in Fig. 6.3(a). Thus, each node is connected to four other nodes in the network. This network indicates a level of clustering of the interactions among nodes, which although common in social networks is often surprising in other kinds of networks that do not involve people.

In social networks, the clustering is measured by a coefficient that determines how interconnected people are. For example, it is far more likely that two friends of mine know each other, than it is for any two individuals randomly selected in the United States to know each other. The clustering coefficient measures how much greater the likelihood of two persons knowing each other is in social networks over what it would be in the network as a whole. This means that in order for a person at the top of the circle in Fig. 6.3(a) to interact with a person at the bottom of the circle, s/he must proceed sequentially through their friends around the circle, being introduced by those they know to those they do not know. So far, there are no shortcuts in this model. However, this circuitous route for interactions is not what is observed in the real world, for if it were, it would take years to expand our circle of friends.

The six degrees of separation phenomenon can be explained by the observation that the above clustering in Fig. 6.3(a) is retained, if a few random connections are included in the circular world depicted by the graph in Fig. 6.3(b). In this "small world", there are short cuts that allow for connections between one tightly clustered group to another tightly clustered group very far away. With relatively few of these long-range random connections, it is possible to link any two randomly chosen individuals with a relatively short path. Consequently, there are two basic elements necessary for a small-world model, clustering and random long-range connections, which can be

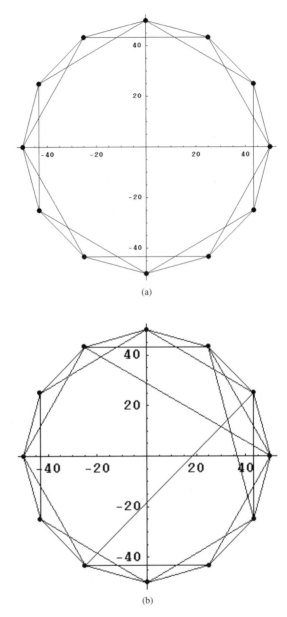

Fig. 6.3: This is a schematic of a small world with clustering, starting from the circle of nodes, as did Duncan Watts and Steven Strogatz. On the top, each node is connected with its nearest and next-nearest neighbor, as indicated, making for a clustered world. To shrink this world a relatively small number of random connections are introduced, as shown on the bottom. These random links produce long-range shortcuts among distant nodes, significantly reducing the average distance between nodes.

interpreted in terms of the concepts of regularity and randomness introduced earlier. This should suggest the connection between the small-world model and the exotic statistics discussed previously.

These are the bare essentials of the Watts-Strogatz small-world model[114] and for a complete discussion and the further development of the network theory, you are encouraged to read the excellent acount of this history by Watts.[111]

6.2. Scale-free networks

Networks that function effectively in the real world do not have the uncorrelated randomness of the Erdös-Rényi network. This difference between idealized mathematical models and the real world has been encountered before. Recall that real physiological time series typically do not have uncorrelated random variations, since there are usually significant connections among the fluctuations. The same is true of complex networks and it is probably easier to understand the nature of the argument concerning correlations using complex networks than the corresponding argument using probability theory, particularly when the technical details have not been mastered.

As mentioned, the Poisson distribution describes the probable number of links that a given node has in a random network. The distribution is based on the assumption that the probability of any new connection is independent of how many connections already exist at a node. This is similar to the flipping of a coin. Each toss is independent of all the preceding tosses, so are the links in a random network. It does not matter how many connections have already been made since each link is a new beginning. This is precisely the mathematical condition that Barabási and Albert[115] determined to be unrealistic, i.e. this independence is not a property of networks in the real world. Using data obtained from Watts and others, Barabási and Albert were able to show that the number of connections between nodes in the real world networks deviate markedly from the distribution of Poisson. In fact, they found that complex networks have the inverse power-law distributions discussed for biological and sociological phenomena in earlier chapters. Furthermore, they recognized the lack of scale in the phenomena described by such inverse power laws and referred to them as *scale-free networks* in keeping with the long history of such phenomena discussed earlier.

Of course scale-free phenomena are not new to the present discussion. They appeared in the works of Pareto in economics, the works of Zipf on natural languages, Auerbach's law on the size of cities, Lotka's law on the number of scientific publications and Willis' discussion of bio-diversity; all verifying that complex systems are often scale-free. What is new in the present discussion are the beginnings of a mathematical language that may provide a way to discuss these things unambiguously. In particular, Barabási and Albert hypothesized two mechanisms leading to the inverse power law in the network context. One of the mechanisms abandons the notion of independence between successive connections within a network. In fact, the mechanism was known in sociology as the *Matthew effect*, since it was taken from the Book of Matthew in the Bible:

> For unto every one that hath shall be given, and he shall have abundance: but from him that hath not shall be taken away even that which he hath.

In more vulgar terms, this is the principle that the rich gets richer and the poor gets poorer. In a network context, this principle implies that the node with the greater number of connections attracts new links more strongly than do nodes with fewer connections, thereby providing a mechanism by which a network can grow as new nodes are added.

The scale-free nature of complex networks affords a single conceptual picture spanning scales from those in the World Wide Web to those within a biological cell. As more nodes are added to the network, the number of links made to existing nodes depend on how many connections already exist. In this way, the oldest nodes, those that have had the most time to establish links, grow preferentially. Thus, some elements in the network have substantially more connections than do the average, many more than predicted by any bell-shaped curve, either that of Poisson or of Gauss. These are the nodes out in the tail of the distribution. In the airplane example, these would be the hubs of Dallas/Fort Worth and Atlanta, and in the sexual liaison example, these would be the don Juans such as the airline stewart Gaton. Power laws are the indicators of such self-organization, so that inverse power laws become ubiquitous, with order emerging from disorder. In the transition from disorder to order, systems gave up their uncorrelated random behavior, characterized by average values and became scale-free, where the system is dominated by critical exponents.

Thus, the two factors, growth and preferential attachment, are responsible for the difference between random network models and real world networks.[109] In each increment of time, a network grows by adding another node and connecting this node to those already present in the network. The connection between the new node and the network is made by preferentially linking to nodes with the greatest number of preexisting connections. The probability of hooking up with a given node is therefore proportional to the number of links that the node already has. The random network with its identical nodes and independent connections has a bell-shaped (Poisson) curve for the distribution in the number of connections, whereas the scale-free network, with its distinctive nodes and preference for popularity, has an inverse power law describing the number of links to a given node in a network. This is analogous to the density-dependence of the statistical measures postulated by Taylor to explain the allometric scaling of biodiversity observed half a century ago.

Whenever a new connection is to be made regardless of the kind of network, the principle of *preferential attachment*[109] comes into play. In deciding which book to read next, selecting among the titles, the fact that one of them is on the *New York Times* bestsellers list could be a deciding factor, or from the famous Yogi Berra, a paralax view of the same effect is:

Nobody goes to that restaurant anymore. It's too crowded.

Preferential attachment is the added importance placed on one node over another because of prior decisions usually made by others. Of course this does not imply that the preferential attachment mechanism is in any way conscious, since inverse power laws appear in networks where the nodes (people) have a choice, and in networks where the nodes (for example, chemicals) do not have a choice, but are strictly governed by physical laws. However, this mechanism implies that the independence assumption, necessary for constructing a random network, is violated by a scale-free network; one described by an inverse power-law distribution. In these latter networks, not all nodes are equivalent. A different strategy leading to this result is given by Lindenberg and West,[116] who show that such inverse power laws are the result of multiplicative fluctuations in the growth process. Such multiplicative fluctuations are, in fact, equivalent to the mechanism of preferential attachment. A random network lives in the forever now, whereas scale-free networks depend on the history of their development.

The question arises as to why scale-free networks appear in such a variety of contexts such as biological, chemical, economic, social, physical and physiological. Similar answers have been found independently by a number of investigators. In the case of the human lung, I determined that a physiological fractal network was preadapted to errors, is unresponsive to random perturbations and therefore has a decided evolutionary advantage, as discussed earlier.[73] The evolutionary advantage would lead one to find inverse power laws expectantly in all manner of biological phenomena, which we do. A decade later, a similar result was found for scale-free networks, in that such networks, e.g. the Internet, were shown to be fairly unresponsive to either random attacks or random failures.[117] An inverse power law in the number of links implies a large number of nodes with relatively few connections, but a non-negligible number of nodes could be considered to be *hubs*. Consequently, random attacks on such networks would most likely destroy or disrupt nodes with only a few connections and therefore have negligible effect. This would be one of the 80% and ignorable according to the *Pareto Principle*. The likelihood of striking a hub at random and thereby having a global effect is relatively small. Therefore, the scale-free network is robust against random attacks. It is this tolerance to local failure that gives rise to the robustness of scale-free networks and this tolerance is a consequence of the architecture of the network. The scale-free network and fractal physiological structures both have an evolutionary advantage because of their fractal properties. On the other hand, such networks are susceptable to attacks directed at the 20%, the hubs, and elimination of only a small number of these hubs would produce catastrophic failure of the system. It is this subsequent strategy that is adopted by terrorists, when these most sensitive elements of the network can be identified and attacked.

6.3. Controlling complexity

Consider the most obvious kind of network in physiology that involve cells. Following the network logic used above, I might consider a biological cell to be the fundamental node in a living system, even though at a higer resolution, the cell itself could be viewed as a network. Consequently, the complex object that is an animal might be viewed as a network of interacting cells, producing a large-scale metabolic reactor. When an animal eats, the food is converted into energy by chemical reactions within the body and these are the interactions among and within the cells. Some energy is stored against an uncertain future. Some energy is used so that the animal can move about and carry out other functions. Finally, some energy goes into heat. The energy generated per unit time (power) by the body, going into each of these categories, is inclusively called the metabolic rate. According to the complex network arguments, the metabolic rate should be related to the total number of chemical reactions (links) occurring per unit time within the body. If the reactions among the cells within the body form a scale-free network, there would be an inverse power law in the number of chemical reactions involving any given cell.

How can I determine empirically if there is any utility to this overly simplified model of how cells contribute to a body's metabolism? One way to test the model is to push the network argument to its logical extreme, even though such extensions are often nonsensical. If the index of the power-law distribution in the number of chemical reactions at each cell is β, this would be the slope of the distribution curve on a doubly logarithmic plot. The total number of reactions is a function of the total number of cells so that the metabolic rate can be expressed in terms of the total number of cells within the body. Proceeding in this vien, since the cell sizes of all the animals are approximately the same, the metabolic rate could be expressed as a function of the total number of cells, which is the total mass of the body. This is a well known result, that being, there is an allometric relation between the metabolic rate of an animal and the animal's total mass (see the mouse-to-elephant curve in Fig. 6.4). The exponent of the mass in the metabolic allometric relation is given by the slope of the best-fit curve and is approximately 0.74.

If I assume that the total number of reactions within the human body is proportionate to the square of the total number of cells, as it would be for a random network dominated by binary chemical reactions, obtained by using the empirical metabolic scaling rate per cell, $4 - 2\beta = -0.26$, therefore I would obtain the approximate value $\beta = 2.13$ for the index of the hypothetical inverse power law of chemical reactions within the body, when this reacting vessel is treated as a scale-free network. This power-law index is consistent with those values obtained for a variety of other scale-free networks in many different contexts, including the computers and routers that

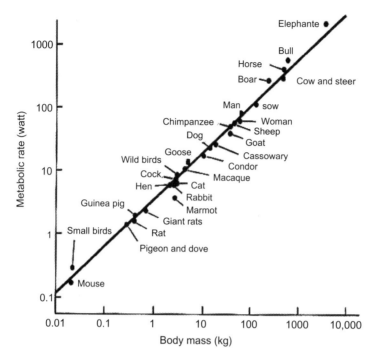

Fig. 6.4: The metabolic rate of selected animals as a function of body mass ranging from a mouse to an elephant.[89] The best-fit straight line gives the metabolic rate as the body mass raised to a power. The power-law index of the mass is obtained from the slope of the solid line and is found to be 0.74.[89]

form the Internet ($\beta = 2.2$),[118] patterns of links among actors ($\beta = 2.3$),[115] the network of coauthors of mathematics papers ($\beta = 2.1$) and neuroscience papers ($\beta = 2.4$) and a long list of other complex phenomena recorded in Tables I and II of Albert and Barabási.[119] It is therefore not crazy to infer that such a scale-free network of chemical reactions by the cells in animals would be consistent with the observed metabolic rates depicted in Fig. 6.4. Note that a direct measurement of the number of reactions within the body yields a power-law distribution with a scaling exponent $\beta = 2.2$, not significantly different from the value of 2.13 obtained using our simple scaling argument. I do not pursue this idea further in this case, as I might if this book were a research monograph, but for our present purposes, I merely use this train of argument to provide additional motivation for applying network ideas to physiology.

Scale-free networks provide an elegant picture of complexity, in which both order and randomness reside, with neither the regularity nor the uncertainty being dominant, just as it was observed with regard to fractals. It is this regular-random duality that gives rise to the robustness of such networks, and may in fact insure their controllability. I need to understand how nature controls scale-free networks without destroying their scale-free property and how that control is related to the allometric nature of the system, i.e. how control is related to the scaling character of the phenomenon.

Recall that the simplest kind of control system samples the output of a network, compares the sampled output with the set point (the desired output) and from the difference between the two, constructs a feedback signal to the input that reduces the separation between the new output and the set point. This is a linear control process that produces a rapid approach of the output towards the set point, by driving the difference between the output and the desired signal to zero. The rate at which the output approaches the set point defines a time scale for the control process. Therefore, this simple negative feedback control cannot be how a scale-free network regulates itself, since this simple control process would destroy the network's scale-free character. I need to devise a control process that is built into the network, does not itself introduce an extraneous scale and still drives the output to the desired form. The notion of a set point might be too limiting for the kind of control I envision for the system-of-systems picture of fractal physiology.

The human body obviously "understands" how to capitalize on the system-of-system architecture of its physiologic networks. The observed scale-free character of these physiologic systems suggests that the control system makes use of the *Pareto Principle*, in which 20% of the nodes carry out most of the functions within the network. A proper control system, one that regulates the system-of-systems response to the environment, would recognize that modification of

80% of the nodes through feedback or otherwise would not materially influence the operation of the network. Only through communication with the significant 20% of the nodes would control be effective. This kind of control would require a feedback mechanism that does not introduce any additional scales into the network, and therefore must itself be scale-free. It is this scale-free property that insures that the control communicates with the right 20% of the nodes. This kind of control emphasizes the uniqueness of the individual nodes and is not democratic.

The system-of-systems architecture determines how multiple complex networks (systems) interact with one another to preform a variety of functions. In particular, this architecture determines how separate and distinct scale-free networks support one another to make the interactive combination more robust than any network acting alone. I previously discussed the cardiorespiratory interaction, in which the slower scale-free breathing is coupled to the faster scale-free beating of the heart. Yet, these systems share a common driver through the sympathetic nervous system, which increases heart rate and breathing rate simultaneously but not necessarily to the same degree. The measure of influence of the external driver is determined by the internal scaling. The key here is that in order for the system-of-systems to retain the robustness of the constitutive systems, it too must be scale-free. This is how the complexity of the cells within our bodies can project their influence upward to macroscopic levels.

The aspirin I take for a headache acts directly at the level of chemistry, that is, on the molecules within my body's cells. Yet, the chemical action produces muscular changes that modify blood flow, which ultimately alleviates pain. Each level in the process communicates with the next level up the chain, but if the initiating step were truly alone, the "voice" of the single chemical reaction would be lost. It is only the chorus of voices at one level that allows it to be heard at the next level. This is what the scale-free network accomplishes. When one of the nodes within the significant 20% obtains information to be transmitted, that node uses the property of scaling, that is, its multiplicity of connections, to amplify the voice of the individual so it can be heard above the din. Consequently, when the aspirin influences one or more of the 20%, the process of pain relief is initiated.

Wiener, the father of cybernetics, distinguished between the kind of control processes associated with voluntary and postural feedback and those asssociated with homeostatic feedback. His contention was that the latter processes tend to be slower than the former. He pointed out that there are very few changes in physiologic homeostasis that produce permanent or even serious damage on time scales of under one second. He goes on to explain that the sympathetic and parasympathetic systems are reserved for the process of homeostasis and the corresponding nerve fibers are often non-myelinated. Such fibers are known to have a considerably slower rate of transmission than do myelinated nerve fibers, and the non-myelinated fibers, although slow, are still the most rapid of the homeostatic transmitters of information. Examples of the non-nervous transmitters include chemical messengers such as hormones and the carbon dioxide content of the blood. The slowing of information transmission, due to the difference between the time scales of the nerve fibers, is compounded by the time scales of the corresponding effectors, see Fig. 2.1. Homeostatic effectors, such as smooth muscles and glands, are slow in their action compared with striated muscle, the typical effectors of voluntary and postural activity. It is apparent that homeostasis is relatively slow, as would be appropriate for the control of an average value, such as heart rate or breathing rate. However, homeostasis would be completely inappropriate for the much faster changes in the fluctuations of the heart rate, breathing rate, stride rate and so forth.

A control process operates on the output of a system in such a way as to bring the system back to the desired behavior (set point), by means of negative feedback. A homeostatic process, as emphasized by Wiener, is typically slow and has a characteristic time scale. So how can such a process be used by a scale-free network? One way is to supplement the simple control process with a complementary allometric control process, which has a broad spectrum of scales contributing to the modification of the system's output. Thus, in the time series of interest, HRV, BRV, SRV, GRV and BTV, the slowly varying homeostatic control is eliminated by taking the difference between quantities. For example, I used the ECG signal and took the difference between successive R peaks to form the HRV time series. This same procedure is used in the other time series as well, where the event-to-event interval, rather than the initial wave form, removes the slow adjustments and focuses the time series on variability. The corresponding time series therefore reveals the behavior of the allometric control process through their scaling.

6.4. Allometric control

So how do I best describe this idea of allometric control and distinguish it from the homeostatic control previously used in the discussion of physiological phenomena? Let us recall the concept of a set point where driving is concerned. The set point is the desired speed and this speed is not set by the driver, but by the municipality. Under the best of conditions, the municipality takes into account how the road winds, the amount of traffic, whether a school is nearby, and other factors that would influence how fast a car should be traveling, before deciding on an appropriate speed limit. The speed limit is posted and in a perfect world, I would modify my driving accordingly. This is not so different from the homeostatic control of the beating heart. Biological evolution, rather than the municipality, has determined the proper rate of blood flow through my body that is required to keep me alive, so that when all the conditions are right, the heart beats at what appears to be a fixed rate with the sympathetic and parasympathetic nerves providing impulses to retain the rhythm.

If I listen to the heart through a stethoscope or feel the pulse at the wrist, the rhythm of the heart seems to be regular. Therefore the cardiac homeostatic control appears to provide a single set point for the beating heart, a dynamic set point where the rate of beating is a constant. However, this is not the case. This dynamic set point is not just a single value, but is instead a distribution of values, all of which are accessible to the cardiovascular control system. So how does the control system choose among the possible values of the interbeat interval? Suppose the rhythmic output of a system is a sinusoidal function. The input to a homeostatic control would be the sinusoid and the control would test the signal to determine if its frequency is contant. One way to test the frequency would be to measure the time interval between successive maxima of the sinusoid and to compare the measured intervals with the set value and adjust the time series accordingly. This of course assumes that the desired output is a single period sinusoidal variation.

However, a physiologic process is open to the environment so that there are always disruptive influences that shift the sinusoid period to larger and smaller values. A homeostatic control might let random (uncorrelated) changes in the period pass, so that the set point would consist of a set of random values. However, by the same token, such control would suppress systematic changes such as a linear ramp that steadily increases or decreases the period. Such a control would retain the average period of the signal. In fact, a homeostatic control might transform any systematic disturbances of the periodic signals into uncorrelated random noise, to which physiologic systems are relatively insensitive.

The linear negative-feedback homeostatic control process typically operates in real-time, so that there is a nearly instantaneous influence of the control on the input to achieve the set point. The rapid nature of the mechanism enables the control to align the input with normal sinus rhythm in heartbeats and with normal stride intervals in real-time walking. Superimposed on this homeostatic control is allometric control. However, rather than being nearly instantaneous, the allometric control has a built-in long-time memory reflected in its inverse power-law character. This memory implies that the allometric control does not only respond to what is happening now, but it also responds to what happened in the past. If the interval from one stride to the next increases, the memory of that increase is carried forward in time for nearly one hundred strides. This is what was found in the stride interval data.[85–87] It cannot be overemphasized that it is a fluctuation in the stride interval that is being remembered for one hundred strides, not the stride interval itself.

Part of the allometric mechanism is that the influence of the fluctuations diminish over time as might be expected. An increase that occurred twenty steps ago or twenty heartbeats ago has less influence on the present stride or heartbeat, than one that occurred on the last step or last heartbeat. The correct way to model this effect is part of the research activity in which a small group of scientists have been engaged over the past two decades. The inverse power-law form of the distribution of intervals between heartbeats indicates the existence of a long-time memory, suggesting that any initially random fluctuations are not allowed to go unnoticed. A fluctuation is duly recorded by the allometric control process and rather than being immediately suppressed as it would be for homeostatic control, it is only slightly suppressed, but mostly remembered to influence the next fluctuation. These rapid variations respond to local changes in the environment and the form of that change is determined by the type of memory contained in the allometric control.

The allometric control communicates a past fluctuation of a given size to the present fluctuations. In order to understand this, let us recap the random walk model where the flipping of a coin determines whether a walker steps to the left or to the right, and put this in the context of homeostatic control. In the coin toss, it is clear that the past is completely separated from the present. It does not matter how many times I have tossed the coin, or how many times a head or a tail has appeared in the sequence of tosses; as I toss the coin now, the outcome is completely unique, independent of its history. This separating of the past from the present is characteristic of independent statistical events. Consequently, the random walk that is determined by the coin tosses is one in which the successive steps are completely independent of one another. This random walk represents the output from a homeostatic control where any systematic influence an environmental perturbation has on the system is broken up by the suppression of organization, so that the disturbances dissolve into uncorrelated random noise to which the system has minimal response. Consequently, a homeostatic control decouples the past from the future in such dynamical systems.

Allometric control acts very differently from homeostatic control. In allometric control, there is long-time memory, so that the coin tosses determining the random walk cannot be independent of one another. This might remind you of d'Alambert and his argument with Bernoulli over the efficacy of smallpox innoculations. The allometric control of a process biases the coin toss. For example, if a head occurs on a given toss, the likelihood of the next outcome being a head is greater than the probability that it will be a tail. In the same way, if a tail occurs on a given toss, the likelihood of it being followed by a tail is greater than the next toss showing a head. This change in the probability outcome is the notion of *persistence* in which a sequence of coin tosses is biased in favor of continuing to do what it has been done in the past; persistence is thus a kind of inertia built into the statistical process. One can introduce a parameter α that measures the degree of bias in this kind of allometric control. For the persistent process, the parameter α is confined to the interval $1 > \alpha > 1/2$, where larger values of the parameter correspond to greater persistence. A value of $\alpha = 1/2$ corresponds to a completely uncorrelated random walk, whereas a value of $\alpha = 1$ corresponds to ballistic motion of a walker moving in a straight line trajectory, like a bullet fired from a gun. Intermediate values of α reveal a combination of regularity and randomness, imposed by allometric control of the input process. Note that neither of the extreme limits is attained by the persistent control.

Persistence is not the only kind of allometric control that is possible. There is the opposite influence that can be expressed by the probability of a head following a head being less than that of a tail following a head. In the same manner, the probability of a head following a tail is greater than the likelihood of a tail following a tail. This kind of allometric control imposes a preference for a change on the random walk rather than persistence and carries the unimaginative name of *antipersistence*. The same parameter is again introduced to measure the degree of antipersistence in the random walk, but now the parameter α is in the interval $1/2 > \alpha > 0$. The closer α is to zero, the greater the potential for change from step to step, where the limit $\alpha = 0$ corresponds to a process that periodically switches back and forth between two values with each step. Neither of the extreme limits is attained by the antipersistent control process.

The simplest kind of allometric control process is therefore one that has the familiar bell-shaped curve for its statistics in the limit where the number of steps becomes very large, but the events are strongly correlated with one another, as measured by the parameter α. These are the persistent and antipersistent time series, whose spectra are inverse power laws. A spectrum indicates the amount of power contained at a particular frequency in a dynamic process. A narrow spectrum, one whose power is focused around a single frequency, is essentially periodic. A fractal process has no such dominant frequency and distributes its power across a broad band of frequencies. As mentioned, one such broadband spectrum has an inverse power law. The negative slope of the inverse power law spectrum generated by the above random walks is given by $(2\alpha - 1)$ and the parameter in the spectrum is the same parameter as that in the random walk. This is one way that the scaling parameter is measured, either from the scaling of the time series or the scaling of the spectrum. This type of correlated random walk process was extensively discussed by Benoit Mandelbrot in a variety of contexts and is now known as *fractional Brownian motion*, at least in the case of infinitely long-time series.

The term *Brownian motion* is used because the type of motion is associated with the fluctuations in the time series observed by the Scottish botonist, Robert Brown, in 1828. He observed that a pollen mote suspended in water did not lie still, but instead move around in an erratic path that he found to be incomprehensible. The observed motion led to a fanciful speculation about a microscopic vital force that animated life. It was not until 1905 that Albert Einstein correctly explained the mysterious movement as being caused by the collisions of the lighter ambient fluid particles with the surface of the larger particle. The instantaneous imbalance in the number of collisions over the surface of the pollen mote produced an imbalance in forces, causing the particle to move in an unpredictable way as determined by the instantaneous direction of the random net force. This phenomenon, now known as diffusion, took the colorful name *Brownian motion* to honor the 19th centuty experimenter.

With regard to this phenomenon, both Einstein and Robert Brown were *reapers*, because the effect named after him was first observed experimentally by the Dutch physician Jan Ingenhousz. The phenomenon was only attributed to Brown because Einstein happened to hear about his work and mentioned Brown's name briefly in his second paper on diffusion, the following year. However, 125 years earlier, in 1785, Ingenhousz sprinkled finely powdered charcoal on an alcohol surface and observed the highly erratic motion of the particles. Like Brown, he was completely mystified as to the origins of the erratic motion of the particles in the fluid. Other accomplishments of Ingenhousz included vaccinating the family of George III of England against smallpox, and observing that sunlight is necessary for the process by which plants grow and purify an atmosphere contaminated by the breathing of animals, i.e. photosynthesis. He powdered his charcoal while he was the court physician to Maria Theresa at an annual salary of 5000 gold gulden.[49]

An uncorrelated random process such as random fluctuations may provide input into an allometric controller, and can then be transformed into a correlated random process as the output, such as in fractional Brownian motion. West *et al.*[120] discuss a formalism that captures the essential features of such statistical processes, but I do not discuss that formalism here because it would require introducing such things as the fractional calculus, something I am not prepared to do. However, I may be able to provide an interesting way to realize such a process in a physiologic context using tools already introduced.

The sympathetic and parasympathetic nervous system both drive heart rate. The former speeds up the heart rate and the latter slows it down. Consequently, there is a continual tug-of-war between these two homeostatic controls, as has been discussed. Thus, the dynamic opposition between these two homeostatic controls is interpreted as antipersistent, because the spectrum obtained from the differences in the HRV time series was found to be inverse power law with a negative slope yielding $\alpha < 1/2$. This model of the change in time interval from beat-to-beat of the human heart in terms of an antipersistent random walk was made by Peng et al.[67] and discussed further by West.[32] This interpretation of the difference between two homeostatic controls naturally lends itself to the picture of an allometric control, where the differences are given the appropriate weightings.

Other physiological fractal phenomena have been modeled as persistent rather than antipersistent random walks, because the power-law index determined by data analysis is greater than one-half. I mentioned earlier that the fractal dimension and this power-law index are related by $D = 2 - \alpha$, so that a persistent process has a fractal dimension in the interval $1 \geq D \geq 1.5$, where $D = 1$ is a completely regular process and $D = 1.5$ is an uncorrelated random process. The HRV, BRV, SRV, GRV and BTV time series all yield fractal dimensions within this interval, as discussed in Chap. 5, and therefore each of these physiological phenomena has both regularity and randomness in their control.

Fractional Brownian motion appears to be the simplest form of allometric control, having as it does statistics given by the law of errors, even though its correlations are inverse power law. The virtue of this model lies in the fact that the errors are not independent, but contain long-term memory that tie events together. This kind of allometric control is analogous to a random network with correlated nodes and is reminiscent of the small-world model.

A somewhat more subtle allometric control is one in which the errors are independent, but the statistics are not given by the bell-shaped curve. In this more general process, it is not the correlation function that is inverse power law, since the correlations are all zero. It is the probability density itelf that is inverse power law or at least has an inverse power-law tail. This kind of allometric control is analogous to the scale-fee network and is represented by a Lévy random walk. In Sec. 4.5, we recall that such a walk has the same distribution of sites as a Lévy-flight, but the length of time required to take a step is dependent on the length of the step.

So how do we construct the more general allometric control that gives rise to scale-free Lévy distributions? Recall the mechanism of Bernoulli scaling from the St. Petersburg Game. In that game, I flipped a coin a number of times and initially placed a bet when the first head would come up in the sequence of tosses. The expected winnings are potentially infinite. In that discussion, I noted that two parameters entered into the processes: (1) the frequency of occurance of an event and (2) the size of the winnings for producing that event. In tossing a coin, the frequency of occurance decreased by a factor of two with each additional head and the size of the winnings increased by a factor of two with each additional head. The increase and decrease nullify each another and the ratio of parameters is one. For a more general process, however, the two parameters do not nullify one another and their ratio determines the fractal dimension of the underlying process.

The latter kind of asymmetric Bernoulli scaling was used to generate a Lévy flight in Chap. 4, Fig. 4.8. The corresponding Lévy control process generates an intermittent output that has clusters, within clusters, within clusters, and subsequently, the probability density of large events occuring in the output of such an allometric control is given by an inverse power-law distribution, i.e. the asymptotic form of a Lévy distribution. Following Peng *et al.*,[67] I used fractal Brownian motion to explain the antipersistence in the HRV differences that they observed and this interpretation was based partly on the spectrum of fluctuations in the difference process. However, the antipersistent interpretation would assign Gaussian statistics to the time series. These statistics are not consistent with those observed in the HRV time series, which they also discovered to be Lévy, the observation made just after Fig. 4.1. Consequently, I would expect this second kind of allometric control, based on Lévy statistics, to be more appropriate for the cardiovascular system than is the first kind, which is based on Gaussian statistics. So how are the two kinds of allometric control reconciled?

The answer to the question of reconciliation lies in the time series generated by a Lévy random walk. The fact that it takes a finite time to complete a step of a given length in a Lévy-walk results in the distribution having a finite width, so that the spectrum for a Lévy-walk does not diverge. The steps themselves are not independent of one another, as they are in the case of the Lévy flight, but scale as they did for fractional Brownian motion. This kind of *fractional Lévy motion* has a scaling that is the ratio of the correlation scaling index $2h$ and the Lévy index μ, that being, $2h/\mu$. The interpretation in terms of an antipersistent process requires that this new scaling index be less than one-half, $2h/\mu < 1/2$, and since the Lévy index is confined to the interval $0 < \mu \leq 2$, I have the condition that $h < 1/2$ a result consistent with the experimental observation of Peng et al.[67]

Note that the Lévy distribution becomes the bell-shaped distribution when $\mu = 2$. In that case, the scaling index for the time series reduces to h, which is identical with the parameter introduced earlier $h = \alpha$. A connection between the predictions from inverse power laws in the correlation functions and those in the statistical distributions is seen here.

6.5. Disease as loss of control

It was a bright sunny day as I drove down the California coast, along the Pacific Coast Highway. I was watching the sun play across the water, only occasionally glancing at the road. Without warning, the road turns back onto itself and I was in a hairpin turn at

50 miles/hour, the tires screeched as I forced the steering wheel away from the shoulder and into the turn. I should have slowed down at a warning sign I barely remembered, but all I could do was to apply the brakes and to hold on to the steering wheel. The backend began to slide out of the turn and nothing responded to my efforts. The car was out of control and I could feel myself skidding towards the sharp drop with the ocean waiting below. I turned the steering wheel in the direction of the skid in the hope of averting near certain death. This is one way to lose control of a system that is guided by simple negative feedback. When the linear homeostatic limits are exceeded, the system goes into wild gyrations and it requires strategies to bring it back within normal range before disaster strikes.

The image of a car going out of control and careening off a cliff is essentially medicine's traditional view of disease. The breakdown or degradation of homeostatic control leads to the failure of a physiological system or network. The effects of neurodegeneration and the infirmities of the elderly are generally interpreted as the deterioration or gradual loss of homeostatic control. Whereas the sudden onset of epileptic siezures or the stopping of the heart are generally interpreted as the abrupt loss of homeostatic control as in the accident above, I do not explore this interpretation of disease further because it is all too familiar, filling most textbooks on the engineering control of physiological networks. I also do not pursue it because I believe it to be a distortion of the fundamental nature of disease. The historical perspective stems from the over-reliance on average measures and the misunderstanding of complexity. The traditional perspective assumes that I can understand complex systems through a knowledge of their average values. However, this is only true of linear additive systems described by the statistics of Gauss. The traditional interpretation is contradicted by the scaling results I have presented and referenced throughout this book. My collegues and me made the following observation a decade ago[121]:

> For at least five decades physicians have interpreted fluctuations in heart rate in terms of the principle of homeostasis: physiological systems normally operate to reduce variability and to maintain

a constancy of internal function. According to this theory... any physiological variable, including heart rate, should return to its "normal" steady state after it has been perturbed. The principle of homeostasis suggests that variations of the heart rate are merely transient responses to a fluctuating environment. One might reasonably postulate that during disease or aging the body is less able to maintain a constant heart rate at rest, so that the magnitude of the variation in heart rate is greater.

A different picture develops when one carefully measures the normal beat-to-beat variations in heart rate... the heart rate may fluctuate considerably even in the absence of fluctuating external stimuli rather than relaxing to a homeostatic, steady state.

Consider the spectra associated with the output of a variety of physiological processes governed by self-similar (fractal) control mechanisms. A sketch of such spectra is given in Fig. 6.5 in which we see a characteristic broadband, inverse power-law dependence of the power contained in a given frequency interval as a function of frequency. The spectrum and the correlation function for a given fractal time series are related by a mathematical transform that has the remarkable property of retaining the power-law form. If the correlation function is a power law in time with index α, then the associated spectrum will also be an inverse power law in frequency with index $-\alpha - 1$ and vice versa. Consequently, if there is a long-term memory, characterized by a scaling property in the correlation function, there is a related scaling property in the spectrum. Goldberger and I speculated almost two decades ago that a variety of pathological disturbances might be associated with alterations in such scaling patterns and that the spectral slope might be of diagnostic and prognostic utility[122]:

> It seems that a fairly universal indicator of pathologic dynamics is a narrowing of the physiologic broadband spectrum, which is indicative of a decrease in healthy variability. This is the content of what we defined as the spectral reserve hypothesis.[123] A loss of variability in cardiac interbeat interval fluctuations, for example, has been noted in a variety of setting for over a quarter of a century,

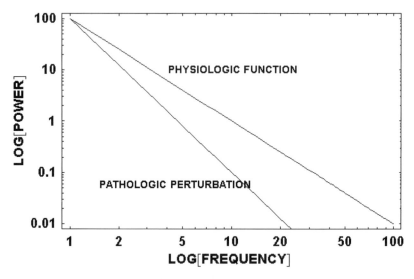

Fig. 6.5: The spectral reserve hypothesis predicts that a variety of apparently unrelated patholic states are characterized by the loss of physiological variability, thus leading to a loss of spectral power. This decrease may be most apparent in high frequency bands. Such a relative decrease in high frequency components would shift the power-law curve resulting in a more negative slope.

including aging,[124] diabetes mellitus and multiple sclerosis[125] and bedrest deconditioning.[126]

In 1987 we speculated that[127]:

> The loss of spectral reserve associated with this decrease in physiologic variability may be of help in quantifying and eventually monitoring response to therapy in a variety of conditions...

This has in fact been proven true. More than a few investigators have shown that changes in the fractal dimension of physiologic time series, the slope of the spectrum, is indicative of pathology. However, it turns out that there is more than one way to lose complexity, as discussed in the next chapter.

There is, however, a technical problem associated with determining the scaling index using the spectrum indicated in Fig. 6.5, having to do with the fact that the properties of physiological time series change over time. This change of statistical properties occurs quite often in physiological phenomena, so that even scaling behavior of the time series can change as increasing data is obtained. This lack of stationarity requires looking beyond traditional spectral methods to determine the scaling behavior of time series. An alternative to the spectral measure of complexity that has some technical advantages in analyzing time series is information, the negentropy that was introduced earlier.

An information measure can be associated with the random time series as an independent test of the scaling behavior of the phenomenon. The procedure is to use the time series to generate a random walk process that allows me to construct a probability density function. This probability density is used to construct the negative entropy or the information associated with the original time series. From the information perspective, it is clear that the underlying dynamics must change over time in order to modify the information generating properties of the dynamical system. Figure 6.6 sketches the increase in information with time, obtained from typical scaling physiological time series. The information increases as I increasingly learn more about a system with each new piece of data, corresponding to an increasing width of the statistical distribution with time. The information increases with the logarithm of time because the underlying process scales and the slope of the information gives the scaling index. When the time series is an uncorrelated Gaussian random process, the slope of the information curve is 0.5. A physical example of such a process is diffusion, obtained, for example, by gently placing a blob of ink in water from an eye dropper. The ink spreads out over space (ambient fluid), giving rise to a variance of the ink blob that increases linearly in time.

The information associated with any particular element of the ink blob increases with the growing uncertainty in the location of that element. This is the condition of maximum entropy for the physical process consistent with the initial state of the system, that being, all the ink concentrated at a single point.

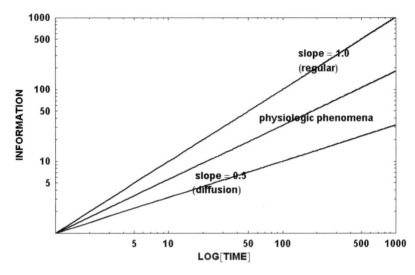

Fig. 6.6: The information generated by the random walk constructed from a physiological time series is plotted versus the logarithm of time. A straight line indicates a scaling realtion relation and the slope of the curve gives the scaling index. The slope of 0.5 corresponds to an uncorrelated random walk with Gaussian statistics. The slope of 1.0 indicates a persistent random walk with ballistic motion. Healthy physiological phenomena register between these two extremes as shown in the preceeding chapter.

The condition of maximum entropy is also interpreted as the phenomenon of maximum disorder, consistent with the initial state of the system. The information gained with any single measurement of the system is therefore also maximum, since it removes the maximum uncertainty in how the process has evolved away from its initial state. Diffusion denotes a process evolving towards the state of equilibrium, which is truly the state of maximum disorder, independently of any initial state of the system. In this physical example of diffusion, equilibrium is reached when the ink is uniformly mixed throughout the ambient fluid.

An equilibrium process in language, such as described by Zipf's law or in scientific publications as described by Lotka's law, or any of the other inverse power-law distributions, yield a certain amount of average information with each new symbol (data point) as discussed earlier. According to Zipf's law, I determined that the average amount of information gained with the addition of each new symbol in a language sequence is 11.82 bits. I can similarly determine that the average information gained with the additon of each new article in a publication sequence is 23.74 bits, since the slope of the Lotka distribution is twice that of the Zipf distribution. However, the information gained with each new data point for a dynamical process is not as simple, since the process changes with time. In particular, if the time series scales, then the amount of information gained with a growing time series is determined by the scaling index, as given by the slope of the curve in Fig. 6.6. In general, the slope of the information curve is the scaling index α, so that a persistent process having $\alpha > 1/2$ produces more information with each new data point, than does a diffusive process with $\alpha = 1/2$. This increase in information gained is associated with the fact that the width of the underlying distribution grows more rapidly in a persistent process than it does during ordinary diffusion. On the other hand, an antipersistent process with $\alpha < 1/2$ produces less information gain per new data point than does a diffusive process.

If the information generated by interbeat intervals, breathing intervals or stride intervals, for normal healthy individuals is plotted, I would obtain straightline curves such as those shown in Fig. 6.6. This is the case because the variability in each of these physiological networks are persistent and generate information faster with each new fluctuation than does diffusion. Furthermore, it is this information gain that I can exploit and determine the state of health of the individual. In Fig. 6.7, I graph the information for a typical stride interval time series and obtain from the slope of the information curve a scaling parameter of 0.69, a value consistent with that obtained using the allometric aggregation technique, see Fig. 5.14. I compare this growth of information with a surrogate data time series constructed from the SRV. The surrogate data is obtained by randomly shuffling the time intervals around in the sequence, neither adding nor subtracting any data. This shuffling of the data destroys any correlation among the data points. Consequently, if the data were

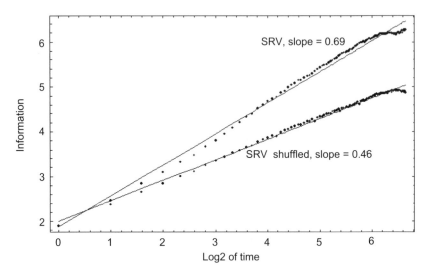

Fig. 6.7: The information generated as a function of time by the stride interval time series or SRV for a typical time series has a slope of 0.69, consistent with the scaling index obtained using the allometric aggregation technique, see Fig. 5.14. This information curve is compared with that generated by the same data shuffled to suppress the long-time correlation.

fractional Brownian motion, we could expect the slope of the information curve to change from 0.7 to 0.5. In other words, upon losing its persistence, SRV would become an uncorrelated random process. This is precisely what is observed, and although 0.46 is not 0.50, it is close enough for a single shuffling of the data. In general, one should construct approximately one hundred surrogates and determine the average scaling parameter for comparison to insure statistical significance.

After shuffling the data points, the information measure did not necessarily have to give a slope of 0.5. It could have given the same slope as that observed before shuffling. The latter result would have meant only one thing; the source of the scaling could not be a correlation among the data points. Instead, the scaling would have been produced by the statistical distribution; the statistics would then be Lévy and not Gaussian. A statistical process with Lévy statistics having an index μ would yield an information curve with a slope given by $1/\mu$, which would be insensitive to shuffling. Consequently, the scaling exponent would be $\alpha = 1/\mu$ and I would have a way of distinguishing between fractional Brownian motion and Lévy statistics as the source of scaling in the data.

It is worthwhile to point out that the implications of the fractal physiology perspective are still being investigated. Only recently, a review[128] focusing on the difference between the classical concepts of physiologic control and fractal control has emphasized the importance of modern statistical physics in understanding the scaling of long-time, inverse power-law correlations.

Chapter 7

Disease as Loss of Complexity

I have argued in a number of places in this book that to understand disease, it is first necessary to set aside the traditional perspective of medical practitioners[19,32,47] regarding health and disease. The first casualty of abandoning the historically accepted view of medicine is the theory of homeostasis, according to which the apparent fluctuations of medical variables such as heart rate are primarily due to external influences, i.e., influences external to the system being investigated. In this accepted view, the normal condition of the cardiovascular and other physiologic systems is that of a steady state in which the rhythmic behavior of the system is sustained under normal operating conditions and variation is in the biological noise. Several lines of evidence presented, including the inverse power-law spectra of the healthy heartbeat, breathing and walking, are consistent with a countervailing theory that chaotic dynamics and fractal time series, not homeostasis, is the "wisdom of the body".[129]

A corollary to the theory of homeostasis, relating health to physiologic constancy, is the notion that disease and other disturbances are likely to cause a loss of regularity. The theory of fractal physiology predicts just the opposite, i.e., a variety of disease states alter autonomic function and lead to a loss of physiological complexity. Consequently, a loss in the variability of the associated time series, as measured by the fractal dimension, and therefore disease, leads to more and not less regularity.[121–123] I present a parade of pathological phenomena in this chapter that show disease to be the loss of complexity and not the disruption of regularity that it has so long been thought to be.

In keeping with this new interpretation of disease, I first discuss physiologic regularity that is indicative of pathology. This regularity is, in fact, periodicity that is completely foreign to the normal operation of the physiologic process being considered.

Next, I discuss certain kinds of cardiovascular disease as exemplars of the loss of complexity as measured by heart rate variability, since that is where a great deal of the work has been done. This is due, at least in part, to the fact that Ary Goldberger is a cardiologist and he has been leading the work on developing this perspective since we first began to collaborate in this area in the early 1980s. In addition to the fact that there is a wide variety of heart diseases against which the new notion of disease can be tested, the HRV time series has also been measured in smokers and the influence of smoking on heart rate can be determined. A final example from the cardiovascular area is the reduction in the multifractality of HRV time series associated with migraine headache.

This last application of the fractal measure as a way to discriminate between a traumatized and a normal fractal time series suggests a totally new use of the scaling index. A person suffering from a traumatic injury and undergoing physical therapy relies on the experience and talent of the therapist to know his/her level of recovery. The scaling index may provide a quantitative indicator of a person's degree of convalescence, without relying solely on the talent of the therapist. The progress of the therapy can also be measured by following the change in the fractal dimension of the physiological phenomenon over time.

The loss of complexity in a phenomenon is evident in stride rate variability as well. Experimental results show that certain neurodegenerative diseases that impair or inhibit walking manifest themselves as a decrease in complexity of the stride rate variability time series. As certain symptoms of individuals with neurodegenerative diseases resemble those in the elderly, the fractal measures of the SRVs of these two groups are compared. Again, the traditional therapy for convalescence using machines such as stationary bicycles and tread mills are probably not the best for a rapid and complete recovery.

The final area of investigation discussed herein has to do with the loss of complexity of breath rate variability. I show some recent results involving ventilation experiments, where it is shown that ventilators with biological variability are vastly superior in facilitating the recovery of the respiratory function, than those with a normal rhythmic driver.

7.1. Pathological periodicities

Adaptability is one of the defining characteristics of health and this is achieved through a broadband spectrum of outputs from the allometric physiologic control systems. The fractal nature of this output implies the lack of any one frequency or scale dominating the dynamics of the physiologic process. The occurance of such periodic behavior would therefore greatly reduce the functional responsiveness of such a system. Consequently, periodicity in the output of a physiological control system is herein interpreted as pathological.

In many severe pathological syndromes, loss of spectral reserve (see Fig. 6.5) is accompanied by a regularity in the underlying time series. This narrowed spectrum, in contrast to the scaling dynamics of the healthy physiological broadband spectrum, is represented by highly periodic behavior. Such pathological periodicities are widespread, not only in cardiovascular dynamics, but in multiple organ systems as well. For example, Cheyne–Stokes syndrome is characterized not only by relatively low-frequency oscillations in respiratory rate and amplitude, but also by periodic modulations

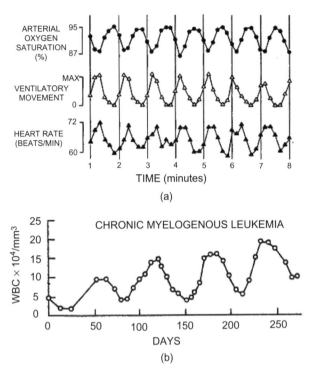

Fig. 7.1: (a) This is a representation of the cyclic pattern in a patient with Cheyne–Stokes breathing, showing waxing and waning periods during which the subject breathes progressively faster and deeper, and then slower and shallower, culimnating in transient complete cessation of breathing. The breathing and heart rate dynamics in a patient with severe heart failure and Cheyne–Stokes breathing. Note the high periodic oscillations of multiple cardiopulmonary variables at the same frequency.[133] (b) Cyclic white blood cell oscillations in a patient with chronic myelogenous leukemia.[134]

of interbeat intervals at the same frequency in some patients (see Fig. 7.1(a)). This type of global oscillatory dynamics, in which an entire system appears to be "enslaved" to a single dominant frequency, stands in stark contrast with physiological oscillations that can be detected as "spikes" riding on the inverse power law spectrum that characterizes healthy variability.

Another example of periodic pathology is taken from hematopoiesis, which is the technical term for the production of blood cells, all of which are generated by the hematopoietic stem cells. The mechanisms regulating the production of blood cells are unclear and an even less well understood phenomenon is the regulation of leukopoiesis, i.e., the production of white blood cells. The regulatory mechanisms in leukopoiesis have been studied using cyclical neutorpenia, a rare periodic hematological disorder characterized by oscillations in the circulating neutrophil count. Scientists are now at a point where they have been able to construct a nonlinear dynamical description of this phenomenon and identify from the model, the potential source of the instabilities producing the oscillations.[130] A related phenomenon is leukemia, a cancer of the white blood cells, which has a number of clinical classifications. Note that certain kinds of chronic leukemia belong to another class of phenomena that shows the loss of variability and the appearance of pathological periodicities. Under normal conditions, the white-blood cell count in healthy individuals appears to fluctuate randomly from day to day, producing an inverse power law. However, in certain cases of chronic leukemia, the white-blood-cell count predictably oscillates periodically (see Fig. 7.1(b)). This dynamic effect is accompanied by the loss of the broadband inverse power-law spectrum and its replacement with a narrowband spectrum, indicative of low-frequency quasi-periodic oscillations.

Other disorders that may manifest themselves through pathological periodicities are epilepsy, Parkinson's disease and manic depression. The motor function in Parkinson's disease can change in a number of different ways, including tremor and rigidity. It is not known exactly why these and other changes occur, but they may be associated with cell destruction in the substantia nigra pars compact region of the brain.[131] Parkinsonian tremor is a large amplitude, low frequency motor oscillation on the order of four to six oscillation per second, occuring as the result of Parkinson's disease. Initially, such uncontrollable tremors were treated by means of surgical procedures to destroy certain regions of the brain, leading to greatly reduced symptoms. It was subsequently determined that an equivalent degree of tremor reduction could be obtained by rapid electrical stimulation of the same region of the brain. There are a number of possible explanations of why deep brain stimulation works to reduce Parkinsonan tremor, but this level of detail is not of concern in this book.[132] This is highlighted as the periodicity that corresponds to disease in the present model and its reduction by a disruption of this periodicity, with a corresponding increase in fractal dimension, are what we interpret as health.

Thus far, I have focused attention on the loss of control associated with the onset of regularity. This type of loss may arise from a rapid transition due to instabilities in the underlying dynamics. A very different loss of control is one in which the spectral reserve is suppressed, but the suppression may happen gradually over time. The maladies one sees with aging are often of this kind. In fact, this kind of loss is not always accompanied by a decrease in fractal dimension, instead, it can arise with an increase in fractal dimension as shown in Fig. 7.2, where we distinguish between the time

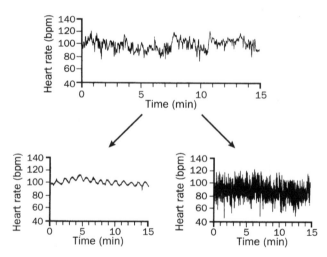

Fig. 7.2: A typical HRV time series with a fractal dimension in the interval $1.1 \leq D \leq 1.3$ is depicted by the upper curve. Two pathology branches emanate from this time series. The one on the left is congestive heart failure with fractal dimension approximately 1.0 and the one on the right is atrial fibrillation with fractal dimension approximately 1.5 (taken from Goldberger[58] with permission).

series for congestive heart failure ($D = 1$) and atrial fibrillation ($D = 1.5$).

7.2. Heart failure and fractal loss

A normal healthy individual's HRV time series has a typical fractal dimension between 1.1 and 1.3,[58] and the fractal dimension obtained using the short time series example presented in Chap. 5 falls within this range. The data is not always used in the form of interbeat intervals, e.g., the data on which the probability density in Fig. 4.7 is based on results from the change in beats per minute, rather than the interbeat intervals themselves. HRV data are shown in Fig. 7.2, which was also taken with permission from the work of Ary Goldberger.

The upper curve in Fig. 7.2 clearly shows the variability in heart rate for a normal healthy adult. The two arrows emanating from the top time series indicate two distinct ways in which the heart can fail to perform its function. On the left, the variation in the heartbeat becomes nearly periodic, or regular, and like the quote from third century Chin Dynasty, the patient will be dead in "four days" (or sooner), unless something is done. This outer left time series is from a patient with congestive heart failure (CHF) and the corresponding fractal dimension is nearly one, which is the fractal dimension of a regular curve. The time series on the right is from a person experiencing atrial fibrillation and the fractal dimension is nearly 1.5, the same as that for an uncorrelated random process. These extremes emphasize two of the dominant pathologies associated with the breakdown of the cardiovascular system. One is the complete loss of randomness, leading to death due to regularity. The other is the complete loss of regularity, leading to death due to randomness. A healthy cardiovascular system cannot survive without the balance of control between the regular and the random; either one alone is apparently fatal.

The two pathologies in Fig. 7.2 indicate two different ways that the cardiac control can lose complexity. Congestive heart failure loses the complexity associated with beat-to-beat interval variability and the heartbeat approaches a constant value so that sinus rhythm becomes truly periodic, which, as was said, is death by monotony. Atrial fibrillation, on the other hand, loses complexity differently. HRV retains the level of variability or even increases the level, such that the cardiovascular system retains its options regarding the length of the time interval between beats, but loses the ability to order these intervals in a useful way. This loss of organization is reflected in the loss of long-range correlations in the HRV time series and results in death by disorganization.

A remarkable thing about the three time series in Fig. 7.2 is that they each have essentially the same average rate. Thus, the physician cannot rely on the patient's pulse and must look for another indicator of pathology. The scaling exponent, or equivalently the fractal dimension, clearly distinguishes among these three types of time series. It must be stressed that the change in scaling indicates the problem but not the solution. The change in scaling does, however, provide an early indicator of something being wrong, whose cause the physician must track down.

7.2.1. Multifractal heartbeats

The scaling property of HRV time series does not remain constant over time for sufficiently long data sets. This means that the scaling index is not the same in the morning and afternoon, and may also change again by evening, independently of a person's level of activity. When a process has more than one fractal dimension (scaling index) over time, the phenomenon is known as multifractal, which is analogous to a simple dynamical system being described by more than one frequency. A frequency (fractal dimension) in simple dynamics is replaced by a spectrum of frequencies (spectrum of fractal dimensions) in complex dynamics (multifractal). It appears that most physiologic phenomena have this multifractal property, so that as the process unfolds, it may present many different fractal dimensions. However, there is a structure that comes with the change. The allometric relation between the variance and average value of the HRV time series introduced earlier must be re-interpreted because of this changing scaling behavior.

An analysis of HRV time series reveals not one but a spectrum of fractal dimensions, or equivalently, a distribution of scaling indices, such as shown in Fig. 7.3. The peak of this spectrum is at the average fractal dimension for the time series and the width of the distribution gives a sense of the degree of variability in the fractal dimension over time for this particular individual. I speculated earlier that the scaling observed in the allometric relation was a consequence of the competition between the sympathetic and parasympathetic nervous systems. This hypothesis was tested by Amaral et al.[135] who were able to apply neuroautonomic blockage and observe the complexity of the resulting HRV signal. They found a decrease in the multifractal complexity as manifested in a reduction of the width of the fractal

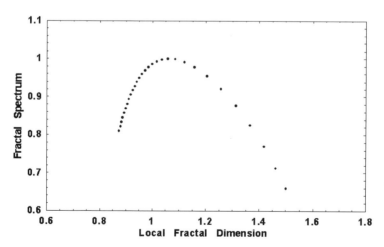

Fig. 7.3: The multifractal spectrum for heartbeat interval (HRV) time series is given at a discrete number of points. This spectrum was calculated using a typical HRV time series from Chap. 5.

spectrum. Consequently, they were able to determine, in part, what supports multifractality in healthy human heartbeat dynamics.

First of all, Amaral et al.[135] determined that reducing the level of external activity for a cohort group of six healthy individuals does not influence the multifractal properties of healthy cardiac dynamics. Secondly, they administered a beta-blocking drug to reduce sympathetic control and determined that the width of the fractal spectrum was greatly reduced, compared with that of a control group. This reduction in width implies that the loss of sympathetic control changes the properties of the HRV time series from multifractal to monofractal, but the time series retains the average fractal dimension of a control group. Consequently, sympathetic blockage may not have a significant effect on the linear correlations of the dynamics. Finally, they administered atrophine to suppress parasympathetic control of the heartbeat and found an even greater reduction in multifractality, than that found with the sympathetic blockage. However, the monofractal dimension, located by the peak of the multifractal spectrum, is shifted to higher values from that of the control group. This shift to a higher dimension indicates a loss of antipersistence in the underlying dynamics. Amaral et al. noted that this result is consistent with the monofractality and the loss of antipersistence observed in heart failure dynamics, and conjectured that impaired parasympathetic control may "explain" congestive heart failure. Bob Lindberg suggested that impaired parasympathetic control is more accurately an association with congestive heart failure, than an explanation for it. Most chronic diseases as well as aging are linked with a reduction in parasympathetic input, and hence reduced HRV. He (Lindberg) sees the increasing sympathetic dominance over the parasympathetic as the body's effort to flog a dying heart, an adaptive mechanism that eventually becomes maladaptive over time. The connection between heart failure and the loss of multifractality was first made by Ivanov et al.,[136] who were also the first group to identify the multifractal character of heartbeat dynamics.

One conclusion that can be drawn from the processing of the data in the above experiments is that the power-law scaling in the allometric relation is not strictly a consequence of the competition between the sympathetic and parasympathetic nervous systems, since the scaling property of the HRV time series persists after neuroautonomic blockage. This persistence suggests that one should look deeper into the dynamics of the separate mechanisms to find complementary causes of their individual fractality. On the other hand, the multifractal character of the time series in a normal healthy individual does appear to be a consequence of this competition. The sympathetic and parasympathetic controls are each separately fractal, but the fractal dimensions of the two are sufficiently far apart, such that when they compete in a normal healthy individual, the result is a multifractal phenomenon.

7.2.2. Effect of smoking

A different kind of complexity loss is observed in a recent study comparing heart rate variability in smokers and non-smokers.[137] The suppression of the high frequency part of the spectrum is observed in a study in which ten young male habitual smokers were matched against an equivalent cohort group of non-smokers. In this limited study, a number of interesting differences between the fractal scaling of the HRV time series of the two groups were determined. The HRV time series was recorded under the conditions of lying and standing quietly to determine the effects of posture on heart rate fluctuations, where both conditions were maintained for half an hour. The first result of interest is that for the non-smokers, there was a significant change in the HRV spectral exponent between the supine and standing positions. In their analysis, the spectrum of fluctuations was determined and the slope of the inverse power-law spectrum was used to determine the scaling exponent. In the supine position, the average slope was approximately -1.0, while the approximate value for standing was −1.3. The increase in slope on standing indicates a loss of high frequency content in the HRV time series, compared with that obtained from the supine position. This effect was not observed in smokers, implying that there was no change in complexity of the HRV time series in the two positions for smokers. However, smokers not only lost this responsiveness to posture, but their spectral slope in both positions was significantly steeper, being approximately −1.45.

Interest in this work does not have to do with the specific values obtained for the scaling index, partly because the spectral technique they used is less reliable than other techniques when applied to non-stationary time series. What was eye catching was the relative change in the values of the scaling indices. Even though I could not rely on the absolute values of the scaling index, I could rely on the observed changes in the scaling index and therefore on the conclusions drawn from these changes. Consequently, habitual smokers show a marked change in the complexity of their HRV time series, indicating a significant modification of the cardiovascular control system from that of the non-smoker "control" group. Esen *et al.*[137] conclude that there is a smoking-related loss of long-range temporal correlation in the HRV allometric regulation, since the spectral exponent was greater in smokers than in age matched non-smokers, independent of posture. They also observe that the steeper slope in the standing position suggests that the upright posture causes a severe decrease in the complexity of autonomic cardiac control, and conclude that this occurs because the upright posture poses challenges to the circulatory homeostasis in humans.

7.2.3. *Effects of aging*

One source of information regarding the fractal physiology interpretation of disease is given on *PhysioNet*, where much of the recent data on heart rates and stride intervals is archieved. In addition, this website has posted an ongoing discussion of how to use modern nonlinear data processing techniques to reveal the often hidden fractal properties of the stored time series data. One area covered by their tutorials has to do with the effects of aging on measured fractal dimensions, in particular, whether alterations in the scaling properties can be of practical diagnostic and prognostic value. One set of experimental data concerns ten healthy young subjects between the ages of 21 and 34, and ten healthy seniors between the ages of 68 and 81. Each subject reclined for two hours in a quiet room, whilst an ECG recorded on a computer their supine resting heart activity.

The data was analyzed using the nonlinear techniques devised by the collaborative group headed by Gene Stanley at Boston University and Ary Goldberger at Harvard Medical School. Their names are constantly mentioned in my discussions because the same small groups have been interested in and working on these problems for a long time; even when we do not collaborate with one another, we track one another's work. Let us recall the random walk exponent introduced earlier. In the healthy young subjects, this scaling exponent takes on a value close to unity, i.e., $\alpha = 1$. The analysis of the senior HRV data, however, did not yield such straightforward results. Rather than a single scaling index, their data showed two scaling regions as sketched in Fig. 7.4.

The two slopes for seniors indicate that the long-time correlation properties are markedly different from those of healthy young adults, evident in the short-time slope. For the youngsters, the unit slope indicates the kind of behavior we saw earlier, i.e., an inverse power-law broadband spectrum implying the long-time correlation determined by the fractal character of the allometric control process. The fractal character of HRV allometric control changes with aging, producing a short-time slope of nearly $\alpha = 1.5$ and an asymptotic slope of $\alpha = 0.5$ for the aged cohort group. These values imply that nearby heartbeats influence one another much more strongly when the subjects were younger, so for short time intervals, there is a loss of versatility in the heart rate with age. The short-time correlation between hearbeats is much greater in the elderly than they are in the young. On the other

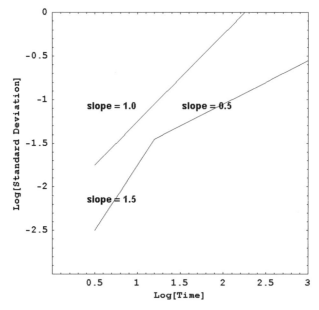

Fig. 7.4: Sketch of the scaling analysis of heartbeat time series in healthy young adults and in healthy seniors. The upper curve with unit slope corresponds to the healthy young adults. The lower curve with two distinct slopes corresponds to the seniors. Here time actually corresponds to the length of the time series used in evaluating the standard deviation.

hand, as the heartbeats become farther apart, there comes a time interval at which any two heartbeats are completely independent of one another. Consequently, the long-time correlations, so evident in the young adults data even at very long times, are absent from the senior data and HRV takes on the character of an uncorrelated random process. In this context, uncorrelated refers asymptotically to the likelihood of a given time interval being followed by a longer time interval, being the same as that followed by a shorter time interval and vice versa.

7.2.4. Heart transplants

Heart transplants is probably one of the most severe tests of the utility of the scaling index as an indicator of health. The complex fluctuations in the healthy heart rate is indicative of a fractal allometric control system and a disruption of this control leads to a loss of complexity in heart rate dynamics. This was interpreted by Khadra et al.[138] as the loss of chaos in the HRV time series, since they associated the fractal dimension with an underlying chaotic dynamics of the heart. Thus, I would expect the statistics of the fluctuations to be random immediately after surgery, when the heart is literally decoupled from the nervous system, but subsequently returning to the signature of the normal healthy individuals, after some of the recipient's nerves grow into the donor heart. Therefore, the method of scaling could be used to detect the return to health as well as any abnormal complications after heart transplantation.[138]

The test Khadra et al. used was to compare the scaling behavior of the HRV time series with 128 surrogates, which they did at different times after the surgery. If the clinical data was indistinguishable from the surrogates, then the time series was random and the patient's new heart had not yet recoupled with the autonomic nervous system. However, when the interbeat intervals changed from being uncorrelated random to being long-time correlated, the patient would be considered to be normal. Of the nine patients tested, seven demonstated that the nervous system had little control on the transplanted heart immediately after the operation, but the control increased and became comparable to that of a normal heart within three months after the operation. The random behavior of the time series in the other two remaining patients persisted.

7.2.5. *Cerebral blood flow*

Migraine headaches have been the bane of humanity for centuries, afflicting notables such as Caesar, Pascal, Kant, Beethoven, Chopin and Napoleon. However, its aetiology and pathomechanism have not been statisfactorily explained to date. Latka et al.[139] demonstrated that the scaling properties of the time series associated with cerebral blood flow (CBF) significantly differs between that of normal healthy individuals and migraineurs. The results of data analysis show that the complex pathophysiology of migraine frequently leads to hyperexcitability of the CBF control systems.

A healthy human brain is perfused by the laminar flow of blood through the cerebral vessels, providing the brain tissue with substrates such as oxygen and glucose. CBF is relatively stable despite variations in systemic pressure and this stability is known as cerebral auto-regulation. Changes in the cerebrovascular resistance of small precapillary brain arteries is a major mechanism responsible for maintaining relatively constant cerebral blood flow. A considerable body of evidence suggests that CBF is influenced by local cerebral metabolic activity. As metabolic activity increases, so does flow, and vice versa. The actual coupling mechanism underlying this metabolic regulation is unknown, but it is most likely to involve certain vasoactive compounds which affect the diameter of cerebral vessels such as adenosine, potassium and prostaglandines, which are locally produced in reponse to metabolic activity. External chemical regulation is predominantly associated with the strong influence of CO_2 on cerebral vessels. An increase in carbon dioxide arterial content leads to marked dilation of vessels (vasolidation) which in turn boosts CBF, while a decrease produces mild vasoconstriction and slows down CBF. The impact of the sympathetic nervous system on CBF is often ignored, but intense sympathetic activity results in vasoconstriction. This type of neurogenic regulation can also indirectly affect CBF via its influence on autoregulation.

Migraine is a prevalent, hemicranial (asymmetric) headache and is among the least understood of diseases. Migraine attacks may involve nausea, vomiting, sensitivity to light, sound, or movement. These associated symptoms distinguish migraine from ordinary tension-type headaches. In about 15% of migraineurs, headache is preceded by one or more focal neurological symptoms, collectively known as the aura. In migraine with aura, the associated symptoms may include transient visual disturbances, marching unilateral paresthesia and numbness or weakness in an extremity, or the face, language disturbances and vertigo. According to the leading hypothesis, migraine results from a dysfunction of brain-stem or diencephalic nuclei that normally modulate sensory input and exert neural influence on cranial vessels. Thus, the fundamental question arises as to whether migraine can significantly influence cerebral hemodynamics. Some experimental data reveal clear interhemispheric blood flow asymmetry in some parts of the brain of migrainerus, even during headache-free intervals.

An indirect way of measuring CBF is by monitoring the blood flowing into the brain through the middle cerebral artery. This can be accomplished using an instrument that operates using the same principle as a radar gun; instead of scattering electromagnetic waves from your car back to the gun to determine your speed, it scatters acoustic (sound) waves from the blood back to the gun to determine the flow velocity. The instrument is a transcranial Doppler ultrasonograph and provides a high resolution measurement of middle cerebral artery flow velocity. We look for the signature of the migraine pathology in the scaling properties of the human middle cerebral artery flow velocity time series. A typical middle cerebral artery flow velocity time series for a healthy subject is depicted in Fig. 7.5.

The dynamical aspects of the cerebral blood flow regulation were recognized by Zhang et al.[140] Rossitti and Stephensen[141] used the relative disperion, the ratio of the standard deviation to the mean,

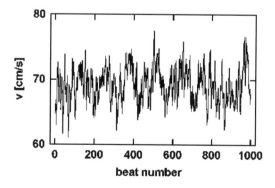

Fig. 7.5: Middle cerebral artery flow velocity time series for a typical healthy subject.[139]

of the middle cerebral artery flow velocity time series to reveal its fractal nature; a technique closely related to allometric aggregation discussed herein. West et al.[142] extended this line of research by taking into account the more general properties of the fractal time series, showing that the beat-to-beat variability in the flow velocity has a long-time memory and is persistent with the average scaling exponent 0.85 ± 0.04, a value consistent with that found earlier in HRV time series. They also observed that cerebral blood flow is multifractal in character.

As mentioned, the properties of monofractals are determined by the local scaling exponent, but there exists a more general class of heterogeneous signals known as multifractals, made up of many interwoven subsets with different local scaling exponents. The statistical properties of these subsets may be characterized by the distribution of fractal dimensions $f(h)$, shown in the multifractal spectrum in HRV time series in Fig. 7.3.

In Fig. 7.6, we compare the multifractal spectrum for the middle cerebral blood flow velocity time series for a healthy group of five subjects and a group of eight migraineurs. A significant change in the multifractal properties of the middle cerebral artery blood flow velocity time series of the migraineurs is apparent. Namely, the interval for the multifractal distribution on the local scaling exponent is vastly constricted. This is reflected in the small value of the width of the multifractal spectrum for the migraineurs, i.e., 0.013, which is almost three time smaller than the width for the control group, i.e., 0.038. For both migraineurs with and without aura, the distributions are centered at 0.81, the same as that of the control group, so the average scaling behavior would appear to be the same. However, the collapse of the fractal dimension spectrum suggests that the underlying process has lost its flexibility. The biological advantage of the multifractal processes is that they are highly adaptive. In the case of a healthy individual, the brain adapts to the multifractality of the interbeat interval time series. We see that the disease, migraine, may be associated with the loss of complexity, and consequently, the

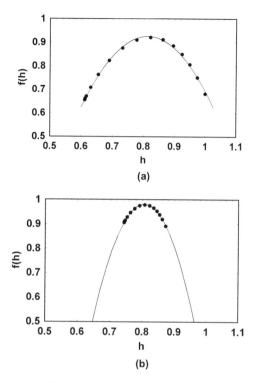

Fig. 7.6: The average multifractal spectrum for middle cerebral blood flow time series is depicted by $f(h)$. (a) The spectrum is the average of 10 time series measurements of five healthy subjects (filled circles). The solid line is the best least-squares fit of the parameters to a predicted spectrum. (b) The spectrum is the average of 14 time series measurements of eight migraineurs (filled circles). The solid curve is the best least-squares fit to a predicted spectrum.[139]

loss of adaptability, thereby suppressing the normal multifractality of cerebral blood flow time series.

Thus, the reduction in the width of the multifractal spectrum is the result of excessive dampening of the cerebral flow fluctuations and is the manifestation of the significant loss of adaptability and overall hyperexcitability of the underlying regulation system. Latka et al.[143] refer to this novel effect as fractal rigidity. They emphasize that hyperexcitability of the CBF control system seems to be physiologically consistent with the reduced activation level of cortical neurons observed in some transcranial magnetic stimulation, and this evoked potential studies.

7.3. Breakdown of gait

The systematic control of complex phenomena such as walking is viewed as allometric, due to the multiplicity of scales involved in the unfolding of the process. Allometric control means a generalization of the idea of feedback regulation that was implicit in the concept of homeostasis. Complex systems producing activities such as gait, involve the elaborate interaction of multiple sensor systems, and have more intricate feedback arrangements than articulated in the original concept of cybernetics. In particular, since each sensor responds to its own characteristic set of frequencies, the feedback control must carry signals appropriate to each of the interacting subsystems. The coordination of the individual responses of the separate subsystems is manifested in the scaling of the time series of the output and the separate subsystems select the aspect of the feedback to which they are the most sensitive. In this way, an allometric control system not only regulates gait, but also readily adapts the stride interval to changing environmental and physiological conditions.

The experimental data on gait, expressed as stride rate variability, is not so extensive as that of the heart rate variability. This is understandable, given that it has not yet been ten years since Hausdorff et al.[84] published the first serious analysis of SRV time series data and established its fractal character using the stride intervals measured in their lab. As mentioned earlier, their gait data is available on PhysioNet and much of it have been used to test the data processing techniques introduced earlier. The influence of aging and neurodegenerative diseases on the SRV time series and how those influences relate to the loss of complexity is of particular interest.

The scaling behavior of HRV time series was observed to change with aging and cardiac pathologies. These results suggest that the same kind of dependences should be observed for gait pathologies reflected in SRV time series. Before discussing the influences of aging and certain neurodegenerative diseases on gait, I examine the influence of stress on SRV time series as determined by Scafetta et al.,[144] before assessing the loss of complexity due to the degradation or breakdown of the motor function.

7.3.1. *Slow, fast and normal gait*

The apparently regular stride experienced during normal walking has random fluctuations with long-time memory that persists over hundreds of steps. As a person changes their average rate of walking, either walking faster or walking slower, the scaling exponent in the fractal time series of the intervals between strides changes with increasing stress. The data on which this analysis was based, was downloaded from the public domain archives of PhysioNet. The data sets are the stride interval time series for ten healthy young men walking at slow, normal and fast paces in both free and metronomically triggered conditions for a period of one hour in the former and thirty minutes in the latter cases, respectively.

The three kinds of data sets, slow, fast and normal, are shown in Fig. 7.7(a) for a typical walker. It is obvious from the figure that the structure of the three stride interval time series is very different for the separate walking conditions. Normal gait is what the walker does after finding a tempo that makes him/her comfortable and which can be maintained for relatively long periods of time. Thus, the middle time series is characteristic of the walker, that being locomotion without stress. When the subjects are asked to walk at the beginning of the trial, faster than normal, or slower than normal, the properties of the time series are visibly different from relaxed normal walking.

It seems to be the case that the slow gait time series is very different from the two other time series. One might suspect that even if normal

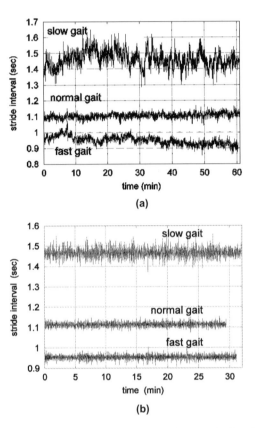

Fig. 7.7: (a) The data sets for three different paces for a typical walker for one hour. (b) The same walker with paces keyed to a metronome for one-half hour. These data were downloaded from PysioNet. (Taken from Ref. 144 with permission.)

gait was a fractal time series, but because the slow gait time series looks so different, the scaling property might be lost. However, this is not the case. The scaling exponent of the normal gait time series averaged over the ten subjects was approximately $\alpha = 0.9$, whereas that for the slow gait time series averaged over the ten subjects was approximately $\alpha = 1.03$, and that for the fast gait is approximately $\alpha = 0.95$. The persistence of the stride interval time series tends to increase for both slow and fast paces, compared with that of the normal pace.

The metronomically constrained walking is depicted in Fig. 7.7(b). The fluctuations for the slow pace remains larger than that for the other two paces, but the wandering that is displayed in the free-pace case has been stabilized by the metronome. When the pace is constrained by the metronome, the stochastic properties of the stride interval time series change significantly. The scaling changes from persistent to antipersistent time series fluctuations. Under each of the three conditions of walking, there is a reduction in the long-term memory and an increase in randomness. This in itself is an interesting result. By keying the system to respond at a set rate, the level of randomness in the allometric control of gait increases rather than decreases. The response to stress is therefore to induce loss of long-time memory.

I noticed in the analysis that some individuals are unable to walk at a set cadence. This inability is recorded in Fig. 7.8 where the multifractal spectra obtained by averaging over the ten healthy young adults is shown for each of the average stride rates. The freely walking time series are evidently multifractal as shown in Fig. 7.8(a) for all three walking rates. The influence of the stress of changing the average rate of one's gait is manifested in the shift in the peak of the multifractal spectrum and a change in the width of the spectrum. On the other hand, this simple picture changes when the individuals attempt to key their steps to an external driver. Their attempts to synchronize their pace to the beat of the metronome resulted in a continual shifting of the stride interval to both longer and shorter strides in the vicinity of the average. For these individuals, the phasing is never right, giving rise to a strong antipersistent signal for all three gait velocities. This is shown by the multiple peaks in the multifractal specta in Fig. 7.8(b) at each of the average rates. These peaks occur because certain individuals are never able to march in formation in the military, nor benefit from any dancing lessons. These individuals cannot adjust to the external driver; they are those with the proverbial two left feet whereby remedy is impossible.

In summary, the stride interval of human gait presents a complex behavior that depends on many factors. Walking is a strongly

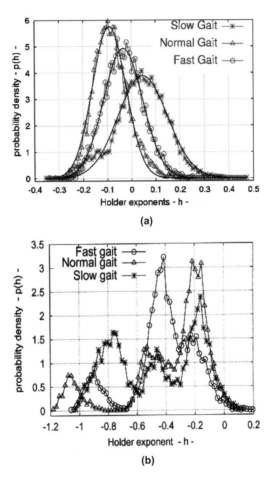

Fig. 7.8: (a) The average multifractal spectrum for ten healthy individuals freely walking at three different rates. (b) The average multifractal spectrum for the same ten individuals as in (a), but constrained by the click of a metronome. (Taken from Ref. 144 with permission.)

correlated neuronal and biomechanical phenomenon which may be greatly influenced by two different stress mechanisms: (i) a natural stress that increases the correlation of the nervous system that regulates the motion at the changing of the gait regime, from a normal relaxed condition to a consciously forced slower or faster gait regime; (ii) a psychological stress due to the constraint of following a fixed external cadence such as a metronome. The metronome causes the breaking of the long-time correlation of the natural pace and generates a large fractal variability of the gait regime.

Hausdorff et al.[145] concluded that the fractal dynamics of walking rhythm are quite robust in general and intrinsic to the locomotor system. Furthermore, the fact that the same people use the same lower motor neurons, both with and without the metronome, actuators and feedback suggest that the metronome is disrupting the usual neurological chain of command. In particular, the metronome is overriding the source of the normal long-time memory in the SRV time series. Therefore, the source of this long-time memory must be at a higher level in the hierarchy than the spinal control. The brain would be one example of such a high level integrator and would also be consistent with our model of allometric control.

Scafetta and I developed a mathematical model of locomotion that governs the stride interval time series for human gait.[146] The model produces a syncopated correlated output associated with a motorcontrol process of the gait cycle. It also incorporates two separate and distinct stress mechanisms. One stress mechanism that has an internal origin increases the correlation of the time series due to a change in the velocity of the gait from normal to the slower or faster regimes. The second stress mechanism with an external origin decreases the long-time correlation of the time series under the frequency constraint of a metronome. We modeled this complex phenomenon by assuming that the intensity of the impulses of the firing neural centers regulate the central frequency of the biomechanical motorcontrol system. Precursor models of motorcontrol were named central pattern generators (CPG), and so with a spark of creativity, we named our model the super central pattern generator (SCPG).[146]

The mathematical model SCPG is able to mimic the complexity of the stride interval sequences of human gait under the several conditions of slow, normal and fast regimes for both walking freely and keeping the beat of the metronome. The SCPG model is based on the assumption that human locomotion is regulated by both the central nervous system and by the motorcontrol system, and is modeled using a nonlinear oscillator with variable frequency. A network of neurons produces a correlated syncopated output (nonlinear oscillator) that is correlated according to the level of psychological stress (external driver), and this network is coupled to the motorcontrol process (variable frequency). The combination of networks controls locomotion as well as the variability of the gait cycle. It is the period of the gait cycle that is measured in the experiments considered above.

7.3.2. Aging and neurodegenerative diseases

The elderly are often prone to falling down without any apparent reasons. Whether this increased clumsiness and loss of grace is a consequence of inattentiveness, or whether it is due to something more basic such as the deterioration of the motorcontrol system, or to a more general neurodegeneration, remains unclear. What is clear is that the scaling of the SRV time series for seniors, when compared with a group of young adults, changes dramatically. The SRV time series for 16 young adults was downloaded from PhysioNet and their scaling properties determined. The scaling exponent was averaged over the cohort group to obtain $\alpha = 0.75$, indicating that these time series are persistent. This is consistent with what was found in Chap. 5, where the example time series had a scaling index of 0.70, which is not the scaling index found in the elderly.

Hausdorff et al.[147] determined that, on average, senior participants and young subjects have virtually indistinguishable average stride intervals and required almost identical lengths of time to perform standardized functional tests of gait and balance. This difference in the measures is not unlike the average heartrates depicted in Fig. 7.2, where the average heart rate ceases to be a useful indicator of cardiac pathology. Using a cohort group of twenty-two healthy young subjects, Hausdorff et al. obtained an average scaling exponent of $\alpha = 0.87 \pm 0.15$, and for ten healthy senior citizens, $\alpha = 0.68 \pm 0.14$, indicating that the stride interval fluctuations are more random and less correlated for the elderly than for the young. This loss of long-time memory arises even though the magnitude of the stride-to-stride variability is very similar for the two groups. Consequently, the gross measures of gait and mobility function do not seem to be affected by age, whereas the scaling exponent clearly changes with age. I wish to exploit this loss of fractality property.

A number of neurophysiological changes occur with age that may alter the locomotor system's ability to correlate fluctuations over time. As Hausdorff *et al.*[148] point out, the physiological changes that occur can include diminished nerve conduction velocity, deafferentiation, loss of motoneurons, decreased reflexes, reduced proprioception and decreased muscle strength, as well as decreased central processing capabilities. These are all the normal effects of aging and may be useful in order to model the subtle changes in neuromuscular control of healthy senior citizens.

Subjects with Huntington's disease, a neurodegenerative disease of the central nervous system, also presented a reduction in the scaling exponent below that of a disease-free control group.[148] The control group of ten subjects was similar to the above young subjects in that they presented a similar scaling index of $\alpha = 0.88 \pm 0.17$, and the group of 17 subjects with Huntington's disease had a scaling similar to the elderly of $\alpha = 0.60 \pm 0.24$. The control and Huntington's groups were chosen to be equivalent to one another, in that they were not statistically different in terms of gender, height or weight, but the latter group was on average slightly older. Note that Hausdorff *et al.* found the severity of Huntington's disease increases inversely with the scaling parameter, i.e., the memory in the fluctuations of the SRV time series decreases with increasing severity of the disease, where the severity is scored using an established measure known as the total functional capacity score of Unified Huntington's Disease Rating Scale. Consequently, the increasing severity of Huntington's disease is measured by the decreasing complexity of the body's gait system.

7.4. Summing up

If one were to form a hierarchy of understanding of scientific complexity, it would surely start with physics as the most basic, expand into chemistry as large aggregates of atoms and molecules form, become biology as life is breathed into these chemical aggregates, and then form physiology, as the phenomenology of human life is explored. Mathematical rigor is demanded at the base of this hierarchy, but mathematical models increase as one climbs the ladder of complexity from physics to physiology. The standard in physics is the sixth place accuracy to which scientists can now theoretically predict the measured value of the fine structure constant. The comfortable mathematical models that physicists rely on are missing in the studies of the biomedical phenomena, and even where such models exist, they do not have the degree of agreement with data that physicists have come to expect. Thus, it might seem to some that the application of modeling concepts such as fractals, scaling, chaotic dynamics, network theory and inverse power-law distributions to physiology is premature. However, unless one is willing to neglect the serious attempts that have been and are being made to apply these ideas to the understanding of physiology and medicine, and dismiss the entire activity as misguided, overviews such as the one given in this book, of how these ideas are being implemented, are useful.

For example, consider the kind of question a physicist might ask in this context. Why is the human body so large relative to the size of atoms? The answer to this question lies in part in the high level of organization necessary for life. The persistence of such organization is possible only in large-scale systems, because macroscopic order would be destroyed by microscopic fluctuations associated with the thermal motion of molecules and biological fluctuations in membranes. Consequently, physiological phenomena have been characterized by averages over ensemble distribution functions generated by microscopic fluctuations. Thus, the strategy for understanding physiology has been based on probabilistic descriptions of complex phenomena involving both deterministic, predictable behavior, and random, unpredictable behavior. When the phenomenon of interest lacks a characteristic scale, then average values are no longer sufficient to characterize it and the variability of the phenomenon ought to be used instead.

The well being of the body's system-of-systems is measured by the fractal scaling properties of the various dynamic subsystems and such scaling determines how well the overall complexity is maintained. Once the perspective that disease is the loss of complexity has been adopted, the strategies presently used in combating disease must be critically examined. Life support equipment is one such strategy, but the tradition of such life support is to supply blood at the average rate of the beating heart, to ventilate the lungs at their average rate and so forth. So how does the new perspective regarding disease influence the traditional approach to healing the body?

Alan Mutch, of the University of Manitoba, has speculated that in analogy with the Internet and other scale-free networks, fractal biological systems are potentially vulnerable to attacks at sites equivalent to network hubs. He also argues that both blood flow and ventilation are delivered in a fractal manner, in both space and time, in a healthy body. However, during critical illness, conventional life support devices deliver respiratory gases by mechanical ventilation, or blood by cardiopulmonary bypass pump in a monotonously periodic fashion. This periodic driving overrides the natural áperiodic operation of the body. Mutch speculates that these devices result in the loss of normal fractal transmission and consequently:[149]

> ...life support systems do more damage the longer they are required and are more problematic the sicker the patient.... We hypothesize that loss of fractal transmission moves the system through a critical point...to transform a cohesive whole to one where organ systems are no longer as well connected.

Disease as the loss of complexity is consistent with the view that complex phenomena have a multiplicity of failure modes. These failure modes result in phenomena-changing character, invariably becoming simpler with an accompanying inability to carry out their functions. A cascade of failures is not so much a consequence of the initiating event, but rather the result of the state of the network when the event is initiated. Consider, for example, an avalanche roaring down the side of a mountain. The avalanche might be triggered by a skier being in the wrong place at the wrong time, or by the explosion of judiciously placed charges set by the ski patrol. In either case, the avalanche is completely out of proportion to the triggering event, resulting from misfortune in the first case and conscious planning in the second. The intentional initiation of small avalanches, for which people are prepared, is the only way to prevent a catastrophic avalanche from occurring when they are not prepared. I adopt such a strategy because science does not understand the dynamic mechanisms by which failures cascade in complex systems, either in a medical or geological setting.

It is in part the irreversibility of failure cascades that makes them so formidable. In medicine, such failure cascades may be manifested as multiple organ dysfunction syndrome (MODS) that rapidly accumulates following a minor insult; MODS is the leading cause of death in intensive care units. As Buchman points out:[150]

> Despite timely and appropriate reversal of the enticing insult... many patients develop the syndrome. Mortality is proportional to the number and depth of system dysfunction and the mortality of MODS after (for example) repair of ruptured abdominal aortic aneurysm is little changed despite three decades of medical progress.

One of the consequences of the traditional view of disease is what Buchman calls "fix-the-number" imperative:[150]

> If the bicarbonate level is low, give bicarbonate; if the urine output is low, administer a diuretic; if the bleeding patient has a sinking blood pressure, make the blood pressure normal. Unfortunately,

such interventions are commonly ineffective and even harmful. For example, sepsis — which is a common predecessor of MODS — is often accompanied by hypocalcaemia. In controlled experimental conditions, administering calcium to normalize the laboratory value increases mortality.

Consequently, one's first choice of options, based on an assumed simple linear causal relationship between input and output as in homeostatsis, is probably wrong.

A number of scientists[151] have demonstrated that the stability of hierarchial biological systems is a consequence of the interactions between the elements of the system. Furthermore, there is an increase in stability resulting from the nesting of systems within systems, i.e., organelles into cells, cells into tissue, tissues into organs etc., from the microscopic to the macroscopic. Each system level confers additional stability on the overall fractal structure. The fractal nature of the system suggests a basic variability in the way systems are coupled together. For example, the interaction between cardiac and respiratory cycles is not constant, but adapts to the physiologic challenges being experienced by the body.

Buchman[150] emphasizes that MODS is not a disease, but rather a syndrome, a usually fatal pathway, yet some patients do get to recover. In such recovery, two features are invariant. Firstly, the time to recover is significantly longer than the time to become ill. This difference in time scales is not unreasonable since MODS is a process by which a cohesive whole is degraded to organ systems that are no longer well connected. The return to health requires the reattachment of the disconnected organs into a cohesive whole again. Life support, lacking fractal delivery patterns, may cause the system as a whole to lose functionality and efficiency; in fact, this lack may inhibit the very reconnections one is trying to establish. The second invariant feature of recovery given by Buchman concerns the fact that measured physiologic parameters do not retrace their paths, implying a metaphorical hysteresis in the clinical trajectory.

A number of scientists have arrived at remarkably similar conclusions regarding the nature of disease that is rather different from the traditional one. The following observation made by Buchman:[150]

> ...Herein, we have suggested that breakdown of network interactions may actually <u>cause</u> disease, and when this breakdown is widespread the clinical manifestation is the mutiple organ dysfunction syndrome. If the hypothesis is correct, then network dysfunction might be expected at multiple levels of granularity, from organ systems to intracellular signal molecules. Restoration of network integrety may be a reasonable therapeutic goal, and a more permissive approach to clinical support (including algorithms that simulate biological variability) might facilitate restoration of network complexity that now appears essential to health.

or those made by Mutch:[149]

> The layer upon layer of fractal redundancy in scale-free biological systems suggests that attack at one level does not place the organism at undue risk. But attack at vital transmission nodes can cause catastrophic failure of the system. The development of multiple organ dysfunction syndrome (MODS) in critically ill humans may be such a failure. Once devolved, death almost inevitably ensues. The similarity to concerted attack on vital Internet router nodes is evident. Patients managed by conventional non-fractal life support may sustain further unintentional attack on a devolving scale-free system due to loss of normal fractal transmission. Returning fractal transmission to life support devices may improve patient care and potentially offer benefit to the sickest of patients.

It is apparent that many of the ideas presented in this book overlap with those articulated by Drs. Buchman and Mutch, even though we approach the matter from the different perspectives of physicists and physicians.

Epilogue

In the prequil, the picture of an acrobat dancing on a wire was developed, moving slow and fast, to keep from falling and to control her motion and locomotion from one side of the circus tent to the other. West and Griffin[89] pointed out that the flowing movements and sharp hesitations of the wirewalker are all part of the spectacle. The wirewalker does not walk in a slow measured step, nor do the physiological systems in the body act in an orchestrated sinus rhythm; both leap and change in unpredicted and unexpected ways to perform their respective functions. This new metaphor is intended to replace the signal plus noise image of the engineer.

For myself, I have reduced what can be known about a physiological system to what I can measure in terms of a time series, whether for the heart, lungs, gait or other system of interest. Such time series have historically been considered to have two complementary properties; those associated with signal and those associated with noise. The signal is taken as the smooth, continuous, predictable, large-scale motion in a process. The notions of signal and predictability go together, in that signal implies information associated with mechanistic, predictable processes in the network. Noise, on the other hand, is typically discontinuous, small-scale, erratic behavior that is seen to disrupt the signal. The noise is assumed, by its nature, to contain no information about the network, but rather to be a manifestation of the influence of the unknown and uncontrollable environment on the system's dynamics. Noise is undesireable and is justifiably filtered out of time series whenever possible.

The signal-plus-noise paradigm constructed by engineers and used by almost everyone to model measurements from complex systems has been replaced in this book with the paradigm of the wirewalker. In the circus, high above the crowd, the tightrope walker carries out smooth average motions plus rapid erratic changes of position, just as in the signal-plus-noise paradigm. However, in the case of the wirewalker, the rapid changes in position are part of the walker's dynamical balance. Far from being noise, these apparently erratic changes in position serve to keep the wirewalker's center of gravity above the wire, so that she does not fall. Thus, both aspects of the time series for the wirewalker's position constitute the signal (both the smooth long-time part and the discontinuous short-time part), and contain information about the wirewalker's dynamics. If I am to understand how the wirewalker retains her balance on the wire, I must analyze the fluctuations, that is, the wirewalker's fine tuning to losses of balance.

Throughout history, heart rate, breathing rate, and steadiness of walking have been used as indicators of health and disease. The research activity of fractal physiology for the past two decades that I have discussed in this book establishes the limitations of these traditional health indicators and proposes new, and more flexible metrics, replacing heart rate with HRV, breathing rate with BRV, and stride interval with SRV. These new metrics suggest the revolutionary view that disease is the loss of complexity. Consequently, I argued that the medical community should relinquish its reliance on averages for understanding wellness and disease, and add measures of fluctuations to its diagnostic repertoire of bedside medicine.

References

1. http://www.halisngegardar.com/eng/historia/histoia2.html.
2. E.W. Montroll and W. W. Badger, *Introduction to the Quantitative Aspects of Social Phenomena*, Gordon & Breach, New York (1974).
3. F.E. Manuel, *A Portrait of Isaac Newton*, pp. 277, A DA CAPO Paperback, Cambridge, MA (1968).
4. J.L. Tarrell, *The Speckled Monster, A Tale of Battling Smallpox*, Dutton, New York (2003).
5. J.M. Fenster, *Mavericks, Miracles and Medicine*, Carrol & Graff, New York (2003).
6. L. Daston, *Classical Probability in the Enlightenment*, Princeton University Press, Princeton, New Jersey (1998).
7. D.E. Smith, *History of Mathematics, Volume 1*, Dover Publication, New York (1958); first published 1923.
8. S. Hollingdale, *Makers of Mathematics*, Penguin Books, London, England (1989).
9. W. Weaver, *Lady Luck*, Dover, New York (1963).
10. C.E. Shannon, "Prediction and Entropy of Printed English", *Bell System Tech. J.* **30**, 50–64 (1951).
11. W.B. Cannon, *The WISDOM of the BODY*, W.W. Norton, NY (1932).
12. http://pespmc1.vub.ac.be/HOMEOSTASIS.html
13. Much of the discussion regarding cybernetics is taken from the encyclopedia article "Cybernetics and Second-Order Cybernetics" by F. Heylighen and C. Joslyn, in *Encyclopedia of Physical Sciences & Technology* (3rd Edition), Academic Press, New York (2001).

14. F. Ashcroft, *Life at the Extremes*, University of California Press, Berkeley (2000).
15. http://medicine.creighton.edu/forpatients/vent/Vent.html.
16. http://asthma.about.com/library/weekly/aa052201a.htm.
17. http://www.johnpowell.net/pages/clevedon.htm.
18. http://www.bae.ncsu.edu/bae/research/blanchard/www/465/textbook/other-projects/1998/respiratory_98/project/history/index.html.
19. B.J. West, *Fractal Physiology and Chaos in Medicine*, World Scientific, Singapore (1990).
20. http://www.si.edu/lemelson/centerpieces/ilives/lecture09.html.
21. N. Wiener, *Time Series*, MIT press, Cambridge, MA (1949).
22. H. Poincare, *The Foundations of Science*, The Science Press, New York (1913).
23. B.J. West, *An Essay on the Importance of Being Nonlinear*, Springer-Verlag, Berlin (1985).
24. Much of the discussion in this subsection comes from B.J. West, Chaos and Related Things: A Tutorial, *The Journal of Mind and Behavior* **18**, 103–126 (1997).
25. R. Thom, *Structural Stability and Morphogenesis*, Benjamin/Cummings, Reading MA (1975).
26. D.W. Thompson, *On Growth and Form* (1915); abridged edition, Cambridge (1961).
27. W.J. Freeman, "Neural mechanisms underlying destabilization of cortex by sensory input", *Physica D* **75**, 151–164 (1994).
28. M. Gladwell, *The Tipping Point*, Little, Brown & Comp., Boston (2000).
29. E.N. Lorenz, *The Essence of Chaos*, University of Washington Press, Seattle (1993).
30. V. Pareto, *Cours d'Economie Politique*, Lausanne (1897).
31. P. Allegrini, M. Giuntoli, P. Grigolini and B.J. West, "From knowledge, knowability and the search for objective randomness to a new vision of complexity", *Chaos, Solitons & Fractals* **20**, 11–32 (2004).
32. B.J. West, *Physiology, Promiscuity and Prophecy at the Millennium: A Tale of Tails*, Studies of Nonlinear Phenomena in Life Science, Vol. 7, World Scientific, New Jersey (1999).
33. G.U. Yule and M.G. Kendal, *Introduction to the Theory of Statistics* (14th Edition), Charles Griffin and Co., London (1950).
34. J.C. Willis, *Age and Area: A Study in Geographical Distribution and Origin of Species*, Cambridge University Press (1922).
35. A.J. Lotka, *Elements of Mathematical Biology*, Dover, New York (1956); original Williams and Wilkins (1924).

36. D. Bickel and B.J. West, "Molecular substitution modeled as an alternating fractal random process in agreement with the dispersion of substitutions in mammalian genes", *J. Molec. Evolution* **47**, 551–556 (1998).
37. F. Auerbach, "Das Gestez der Bebolkerungskonzentration", Petermanns Mittielungen (1913).
38. L.A. Adamic, "Zipf, Power-laws, and Pareto — a ranking tutorial", http://www.hpl.hp.com/research/idl/papers/ranking/ranking.html.
39. G.K. Zipf, *Human Behavior and The Principle of Least Effort*, Addison-Wesley Press. Cambridge (1949).
40. L. Brillouin, *Science and Information Theory*, 2nd Edition, Academic Press, New York (1962).
41. http://www.gospel.mcmail.com/a-g/dry_bones.htm.
42. T.M. Proter, *The Rise of Statistical Thinking 1820–1900*, Princeton University Press, New York (1986).
43. F. Liljeros, C.R. Edling, L.A. Nunes Amaral, H.E. Stanley and Y. Aberg, "The web of human sexual contacts", *Nature* **411**, 907 (2001).
44. R.M. Anderson and R.M. May, "Epidemiological parameters of HIV transmission", *Nature* **333**, 514 (1988).
45. R. Shilts, *And the Band Played On*, St. Martins Press, New York (1987).
46. M. Loève, "Paul Levy's scientific work", *Ann. Prob.* **1**, 1–8 (1973).
47. J.B. Bassingthwaighte, L.S. Liebovitch and B.J. West, *Fractal Physiology*, Oxford University Press, New York (1994).
48. J. Klafter, G. Zumofen and M.F. Shlesinger, "Lévy Description of Anomalous Diffusion in Dynamical Systems", in *Lévy Flights and Related Topics in Physics*, Eds. M.F. Shlesinger, G.M. Zaslavsky and U. Frisch, Springer, Berlin (1994).
49. E.W. Montroll and B.J. West, "On an enriched collection of stochastic processes", in *Fluctuation Phenomena*, Eds. E.W. Montroll and J.L. Lebowitz, North-Holland, Amsterdam (1979); 2nd Edition ibid. (1987).
50. B.J. West and A.L. Goldberger, "Physiology in Fractal Dimensions", *American Scientist* **75**, 354–64 (1987).
51. A.L. Goldberger, V. Bhargava, B.J. West and A.J. Mandell, "On a mechanism of cardiac electrical stability: the fractal hypothesis", *Biophys. J.* **48**, 525–528 (1985).
52. S.K.S. Huard and M. Wong, Chinesische Medizin (aus dem Französischen von H.W.A. Schoeller) Kindler, München (1991).
53. A.L. Goldberger and B.J. West, "Chaos in physiology: health or disease?", in *Chaos in Biological Systems*, 1–5, Eds. A. Holton and L.F. Olsen, Plenum (1987).

54. L. Glass, M.R. Guevara and R. Perez, "Bifurcation and chaos in a periodically stimulated cardiac oscillator", *Physica D* **7**, 39–101 (1983).
55. A.L. Goldberger and B.J. West, "Applications of nonlinear dynamics to clinical cardiology", in *Perspectives in Biological Dynamics and Theoretical Medicine*, Ann.N.Y. Acad. Sci. **504**, 195–215 (1987).
56. A.L. Goldberger and B.J. West, "Fractals; a contemporary mathematical concept with applications to physiology and medicine", *Yale J. Biol. Med.* **60**, 104–119 (1987).
57. Benoit B. Mandelbrot, *Fractals, Form and Chance*, W.H. Freeman, San Francisco (1977).
58. A. Goldberger, "Non-linear dynamics for clinicians: chaos theory, fractals and complexity at the bedside", *The Lancet* **347**, 1312–1314 (1996).
59. L.R. Taylor, "Aggregation, variance and the mean", *Nature* **189**, 732–735 (1961).
60. L.C. Hayek and M.A. Buzas, *Surveying Natural Populations*, Columbia University Press, New York (1997).
61. L.R. Taylor and R.A.J. Taylor, "Aggregation, migration and population mechanics", *Nature* **265**, 415–421 (1977).
62. J.S. Huxley, *Problems of Relative Growth*, Dial Press, New York (1931).
63. L.R. Taylor and I.P. Woiwod, "Temporal stability as a density-dependent species characteristic", *J. Animal Ecol.* **49**, 209–224 (1980).
64. "Heart rate variability", *European Heart Journal* **17**, 354–381 (1996).
65. M.C. Teich, S.B. Lowen, B.M. Jost and K. Vive-Rheymer, "Heart Rate Variability: Measures and Models", arXiv: physics/0008016v1 (2000).
66. B. Suki, A.M. Alencar, U. Frey, P.Ch. Ivanov, S.V. Buldyrev, A. Majumdar, H.E. Stanley, C.A. Dawson, G.S. Krenz and M. Mishima, "Fluctuations, Noise and Scaling in the Cardio-Pulmonary System", *Fluctuations and Noise Letters* **3**, R1–R25 (2003).
67. C.K. Peng, J. Mistus, J.M. Hausdorff, S. Havlin, H.E. Stanley and A.L. Goldberger, "Long-range anticorrelation and non-Gaussian behavior of the heartbeat", *Phys. Rev. Lett.* **70**, 1343–46 (1993).
68. R. Karasik, N. Sapir, Y. Ashkenazy, P.Ch. Ivanov, I. Dvir, P. Lavie and S. Havlin, "Correlation differences in heartbeat fluctuations during rest and exercise", *Phys. Rev. E* **66**, 062902 (2002).
69. R.G. Turcott and M.C. Teich, "Fractal character of the electrocardiogram: Distinguishing heart-failure and normal patients", *Ann. Biomed. Eng.* **24**, 269–293 (1996).

70. B.J. West, W. Zhang and H.J. Mackey, "Chaos, Noise and Biological Data", in *Fractals in Biology and Medicine*, Eds. T.F. Nonnenmacher, G.A. Losa and E.R. Weibel, Birkhäuser (1994).
71. G. Sugaihara, W. Allan, D. Sobel and K.D. Allan, "Nonlinear control of heart rate variability in human infants", *Proc. Nat. Acad. Sci. USA* **93**, 2608–2613 (1996).
72. J. Hilderbrandt and A.C. Young, "Anatomy and physics of respiration" in *Physiology and Biophysics*, 19th ed., Editors T.C. Ruch and H.D. Patton, Saunders, Philadelphia (1966).
73. B.J. West, "Physiology in fractal dimension: error tolerance", *Ann. Biomed. Eng.* **18**, 135–149 (1990).
74. H.H. Szeto, P.Y. Cheng, J.A. Decena, Y. Chen, Y. Wu and G. Dwyer, "Fractal properties of fetal breathing dynamics", *Am. J. Physiol.* **262** (Regulatory Integrative Comp. Physiol. 32) R141–R147 (1992).
75. G.S. Dawes, H.E. Fox, M.B. Leduc, E.C. Liggis and R.T. Richards, "Respiratory movement and rapid eye movements in the fetal lamb", *J. Physiol. Lond.* **220**, 119–143 (1972).
76. B.J. West, L.A. Griffin, H.J. Frederick and R.E. Moon, "The Independently Fractal Nature of Respiration and Heart Rate During Exercise Under Normobaric and Hyperbaric Conditions", *Respiratory Physiology & Neurobiology* **145**, 219–233 (2005).
77. R. Karasik, N. Sapir, Y. Ashenazy, P.C. Ivnaov, I. Dvir, P. Lavie and S. Havlin, "Correlation differences in heartbeat fluctuations during rest and exercise", *Phys. Rev. E* **66**: 062902(4) (2002).
78. W.A.C. Mutch, S.H. Harm, G.R. Lefevre, M.R. Graham, L.G. Girling and S.E. Kowalski, "Biologically variable ventilation increases arterial oxygenation over that seen with positive end-expiratory pressure alone in a porcine model of acute respiratory distress syndrome", *Crit. Care Med.* **28**, 2457–64 (2000).
79. W.A. Altemeier, S. McKinney and R.W. Glenny, "Fractal nature of regional ventilation distribution", *J. Appl. Physiol.* **88**, 1551–1557 (2000).
80. C.K. Peng, J. Metus, Y. Li, C. Lee, J.M. Hausdorff, H.E. Stanley, A.L. Goldberger and L.A. Lipsitz, "Quantifying fractal dynamics of human respiration: age and gender effects", *Ann. Biom. Eng.* **30**, 683–692 (2002).
81. J.W. Kantelhardt, T. Penzel, S. Rostig, H.F. Becker, S. Havlin and A. Bunde, "Breathing during REM and non-REM sleep: correlated versus uncorrelated behaviour", *Physica A* **319**, 447–457 (2003).

82. C.L. Webber, "Rhythmogenesis of deterministic breathing patterns", in *Rhythms in Physiological Systems*, pp. 177–191, Eds. H. Haken and H.P. Koepchen, Springer, Berlin (1991).
83. Vierordt, Ueber das Gehen des Menchen in Gesunden und kranken Zustaenden nach Selbstregistrirender Methoden, Tuebigen, Germany (1881).
84. J.M. Hausdorff, C.-K. Peng, Z. Ladin, J.Y. Ladin, J.Y. Wei and A. L. Goldberger, "Is walking a random walk? Evidence for long-range correlations in stride interval of human gait", *J. Appl. Physiol.* 78(1), 349–358 (1995).
85. B.J. West and L. Griffin, "Allometric control of human gait", *Fractals* 6, 101–108 (1998).
86. B.J. West and L. Griffin, "Allometric Control, Inverse Power Laws and Human Gait", *Chaos, Solitons & Fractals* 10, 1519–1527 (1999).
87. L. Griffin, D.J. West and B.J. West, "Random Stride Intervals with Memory", *J. Biol. Phys.* 26, 185–202 (2000).
88. R.L. Marsh, D.J. Ellerby, J.A. Carr, H.T. Henry and C.I. Buchanan, "Partitioning the Energetics of Walking and Running: Swinging the Limbs is Expensive", *Science* 303, January (2004).
89. B.J. West and L. Griffin, *Biodynamics: Why the Wirewalker Doesn't Fall*, Wiley & Sons, New York (2004).
90. J.M. Hausdorff, L. Zemany, C.-K. Peng and A.L. Goldberger, "Maturation of gait dynamics: stride-to-stride variability and its temporal organization in children", *J. Appl. Physiol.* 86, 1040–1047 (1999).
91. J.J. Collins and C.J. De Lucca, "Random walking during quiet standing", *Phys. Rev. Lett.* 73, 764–767 (1994).
92. J.W. Blaszczyk and W. Klonowski, "Postural stability and fractal dynamics", *Acta Neurobiol. Exp.* 61, 105–112 (2001).
93. B.J. West, A. Maciejewski, M. Latka, T. Sebzda and Z. Swierczynski, "Wavelet analysis of scaling properties of gastric electrical activity", *J. Appl. Physiol.*
94. E.R. Kandel, "Small systems of neurons", *Sci. Am.* 241, 66–76 (1979).
95. G.L. Gernstein and B. Mandelbrot, "Random walk models for the spike activity of a single neuron", *Biophys. J.* 4, 41–68 (1968).
96. M.C. Teich, C. Heneghan, S.B. Lowen, T. Ozaki and E. Kaplan, "Fractal character of the neural spike train in the visual system of the cat", *Opt. Soc. Am.* 14, 529–546 (1997).
97. A. Babloyantz and A. Destexhe, "Low-dimensional chaos in an instance of epilepsy", *Proc. Nat. Acad. Sci. USA* 83, 3515–3517 (1987).
98. http://www.physionet.org/tutorials/fmnc/node16.html.
99. J.A. Hirsch and B. Bishop, "Respiratory sinus arrhythmia in humans; how breathing pattern modulates heart rate". *Am. J. Physiol.* 241, H620–H629 (1981).

100. P. Calabrese, H. Perrault, T.P. Dinh, A. Eberhard and G. Benchetrit, "Cardiorespiratory interactions during resistive load breathing", *Am. J. Physiol. Reg. Int. Comp. Physiol.* **270**, R2208–R2213 (2000).
101. C.K. Peng, J. Metus, Y. Liu, G. Khalsa, P.S. Douglas, H. Benson and A.L. Goldberger, "Exaggerated Heart Rate Oscillations During Two Meditation Techniques", *Int. J. Cardio.* **70**, 101–107 (1999).
102. J.E. Mietus, C.K. Peng, P.C. Ivanov and A.L. Goldberger, "Detection of Obstructive Sleep Apnea from Cardiac Interbeat Interval Time Series", *Comp. in Cardio.* **27**, 753–756 (2000).
103. D.C. Galletly and P.D. Larsen, "Cardioventailatory coupling in heart rate variability: methods for qualitative and quantitative determination", *B. J. Anaes.* **87**, 827–833 (2001); "Cardioventilatory coupling in heart rate variability: the value of standard analytical techniques", *B. J. Anaes.* **87**, 819–826 (2001).
104. Y. Yamamot, J. Fortrat and R.L. Hughson, "On the fractal nature of heart rate variability in humans: effects of respiratory sinus arrhythmia", *Am. J. Physiol.* **269** (*Heat Circ. Physiol.* **38**): H480–H486 (1995).
105. W. Rall, "Theory of physiological properties of dendrites", *Annals of New York Academy of Science* **96**, 1071–1091 (1959).
106. C.D. Murray, "A relationship between circumference and weight and its bearing on branching angles", *Journal General Physiology* **10**, 125–729 (1927).
107. J.P. Richter, Editor, *The Notebooks of Leonardo da Vinci*, Vol. 1, Dover, New York; unabridged edition of the work first published in London in 1883.
108. N. MacDonald, *Trees and Networks in Biological Models*, John Wiley & Sons, New York (1983).
109. M. Buchanan, *Nexus*, W.W. Norton & Co., New York (2002).
110. A. Barabási, *Linked*, Plume, New York (2003).
111. D.J. Watts, *Six Degrees*, W.W. Norton & Co., New York (2003).
112. *Statistical Mechanics and Statistical Methods in Theory and Application: A Tribute to Elliott W. Montroll*, see page 784, Ed. U. Landman, Plenum Press, New York (1977).
113. M. Granovetter, "The strength of weak ties", *American Journal of Sociology* **78**, 1360–1380 (1973).
114. D.J. Watts and S.H. Strogatz, "Collective dynamics of 'small-world' networks", *Nature* **393**, 440–442 (1998).
115. A.-L. Barabasi and R. Albert, "Emergence of scaling in random networks", *Science* **286**, 173 (1999).
116. K. Lindenberg and B.J. West, *The Nonequilibrium Statistical Mechanics of Open and Closed Systems*, VCH, New York (1990).
117. R. Albert, H. Jeong and A. Barabasi, "Attack and Error Tolerance of Complex Networks", *Nature* **406**, 378 (2000).

118. M. Flautsos, P. Flautsos and C. Flautsos, "On power-law relationships of the Internet Topology", *Computer Communication Review* **29**, 251 (1999).
119. R. Albert and A.-L. Barbasi, "Statistical Mechanics of Complex Networks", *Reviews of Modern Physics* **74**, 47 (2002).
120. B.J. West, M. Bologna and P. Grigolini, *The Physics of Fractal Operators*, Springer, New York (2003).
121. A.L. Goldberger, D.R. Rigney and B.J. West, "Chaos and Fractals in Human Physiology", *Scientific American*, February, 42–49 (1990).
122. B.J. West and A.L. Goldberger, "Physiology in Fractal Dimensions", *American Scientist* **75**, 354–365 (1987).
123. A.L. Goldberger, B.J. West and V. Bhargava, "Nonlinear mechanisms in physiology and pathophysiolgy. Toward a dynamical theory of health and disease. "*Proceedings of the 11th IMACS World Congress, Oslo, Norway,* Vol. 2, Eds. B. Wahlsrom, R. Henriksen, and N.P. Sundby, 239–243, North-Holland, Amsterdam (1985).
124. J.L. Waddington, M.J. MacColloch and J.E. Sambrooks, *Experientia* **35**, 1197 (1979).
125. N. Neubauer and H.J.G. Gundersen, *J. Neurol. Neurosurg. Psychiatry* **41**, 417 (1978).
126. A.L. Goldberger, L.D. Goldwater and V. Bhargava, *J. Appl. Physiol.* **61**, 1843 (1986).
127. A.L. Goldberger and B.J. West, "Applications of Nonlinear Dynamics to Clinical Cardiology", in *Perspectives in Biological Dynamics and Theoretical Medicine*, Annals New York Academy of Sciences **504**, 195–213 (1987).
128. A.L. Goldberger, L.A.N. Amaral, J.M. Hausdorff, P.Ch. Ivanov, C.-K. Peng and H. Eugene Stanley, "Fractal dynamics in physiology: Alterations with disease and aging", *PNAS* **99**, February 19 (2002).
129. A.L. Goldberger and B.J. West, "Chaos and order in the human body", *MD Computing* **9**, 545–554 (1992).
130. S. Bernard, J. Belair and M.C. Mackey, "Oscillations in cyclical neutropenia: new evidence based on mathematical modeling", *Journal of Theoretical Biology* **223**, 283–298 (2003).
131. R. Edwards, A. Beuter and L. Glass, "Parkinsonian Tremor and Simplification in Network Dynamics", *Bull. Math. Biol.* **51**, 157–177 (1999).
132. M.S. Titcombe, L. Glass, D. Guehl and A. Beuter, "Dynamics of Parkinsonian tremor during deep brain stimulation", *Chaos* **11**, 766–773 (2001).
133. L.J. Findley, M.R. Blackburn, A.L. Goldberger and A.J. Mandell, *Am. Rev. Resp. Dis.* **130**, 937–939 (1984).

134. A.L. Goldberger, "Fractal variability versus pathologic periodicity: complexity loss and stereotypy in disease", *Perspecives in Biology* **40**, 543–561 (1997).
135. L.A.N. Amaral, P.Ch. Ivanov, N. Aoyagi, I. Hidaka, S. Tomono, A.L. Goldberger, H.E. Stanley and Y. Yamamoto, "Behavior-Independent features of complex heartbeat dynamics", *Phys. Rev. Lett.* **86**, 6026–6029 (2001).
136. P.Ch. Ivanov, L.A.N. Amaral, A.L. Goldberger, S. Havlin, M.G. Rosenblum, A.R. Struzik and H.E. Stanley, "Multifractality in human heartbeat dynamics", *Nature* **399**, 461–465 (1999).
137. F. Esen, F.Y. Ozbebit and H. Esen, "Fractal Scaling of Heart Rate Variability in Young Habitual Smokers", *Truk. Med. Sci.* **31**, 317–322 (2001).
138. L.M. Khadra, T.J. Maaya and H. Dickhaus, "Detecting Chaos in HRV Signals in Human Cardiac Transplant Recipients", *Computers and Biomedical Research* **30**, 188–199 (1997).
139. B.J. West, M. Latka, M. Glaubic-Latka and D. Latka, "Multifractality of cerebral blood flow", *Physica A* **318**, 453–460 (2003).
140. R. Zhang, J.H. Zuckerman, C. Giller and B.D. Levine, "Spontaneous fluctuations in cerebral blood flow: insights from extended-duration recordings in humans", *Am. J. Physiol. Heart Circ. Physiol.* **278**, 1848–1855 (2000).
141. S. Rossitti and H. Stephensen, "Temporal heterogeneity of the blood flow velocity at the middle cerebral artery in the normal human characterized by fractal analysis", *Acta Physio. Scand.* **151**, 191 (1994).
142. B.J. West, R. Zhang, A.W. Sanders, J.H. Zuckerman and B.D. Levine, "Fractal Fluctuations in Cardiac Time Series", *Phys. Rev. E* **59**, 3492 (1999).
143. M. Latka, M. Glaubic-Latka, D. Latka and B.J. West, "Fractal Rididity in Migraine", *Chaos, Solitons & Fractals* **20**, 165–170 (2003).
144. N. Scafetta, L. Griffin and Bruce J. West, "Holder exponent spectra for human gait", *Physica A* **328**, 561–583 (2003).
145. J.M. Hausdorff, Y. Ashkenazy, C.-K. Peng, P.Ch. Ivanov, H.E. Stanley and A.L. Goldberger, "When human walking becomes random walking: fractal analysis and modeling of gait rhythm fluctuations", *Physica A* **302**, 138–147 (2001).
146. B.J. West and Nicola Scafetta, "Nonlinear dynamical model of human gait", *Phys. Rev. E* **67**, 051917–1 (2003).
147. J.M. Hausdorff, C.K. Peng, J.Y. Wei and A.L. Goldberger, "Fractal Analysis of Human Walking Rhythm", in *Biomechanics and Neural Control of Posture and Movement*, J.M. Winters and P.E. Crago (eds.) Springer, 2000.
148. J.M. Hausdorff, S.L. Mitchell, R. Firtion, C.K. Peng, M.E. Cudkiowicz, J.Y. Wei and A.L. Goldberger, "Altered fractal dynamics of gait: reduced stride-interval

correlations with aging and Huntington's disease", *J. Appl. Physiol.* **82**, 262–269 (1997).
149. A. Mutch, "Health, 'Small-Worlds', Fractals and Complex Networks: An Emerging Field", *Med. Sci. Monit.* **9**, MT55–MT59 (2003).
150. T.G. Buchman, "Physiologic failure: multiple organ dysfunction syndrome", in *Complex Systems Science in BioMedicine*, T.S. Deisboeck and S.A. Kauffman, (eds.) Kluwer Academic-Plenum Publishers, New York (2006).
151. G.A. Chauvet, "Hierarchial functional organization of formal biological systems: a dynamical approach. I. The increase of complexity by self-association increases the doman of stability of a biological system", *Philos. Trans. R. Soc. Lond. B Biol. Sci.* **339**(1290), 425–444 (1993).

Index

afferent pathway, 42
allometric relation, 186
antidisease, 136
antipersistence, 269
attractor, 103
Bernoulli's Theorem, 19
central limit theorem, 85
complex adaptive systems, 44
controller, 42
creepers, 124
cybernetics, 44
effectors, 43
efferent pathway, 43
feedback, 43
feedforward, 44
fractal physiology, 1, 68, 114, 175
fractal time, 182
fractional Brownian motion, 269
fractional Lévy motion, 273
goose bumps, 49
hubs, 259
keepers, 124
Kinsey Report, 156
Law of Large Numbers, 19
leapers, 124
March of Dimes, 51
Matthew effect, 256
negative feedback, 43
Pareto Principle, 147, 170

perfect gas law, 74
persistence, 268
Physionet, 159
positive feedback, 43
preadapted, 208
preferential attachment, 257
prohealth, 136
Quincunx, 84
reaper, 124, 270
redundancy, 171
Salk Institute, 131
St. Petersburg paradox, 164
scale-free networks, 256
set point, 41
sleepers, 124
The Law of Frequency of Errors, 77

action and reaction, 75
Adamic, 150
adaptive, 300
ADD, 17
aether, 73
aggregate, 187, 310
aging, 294
AIDS, 158
airports, 247
airways, 205
Albert, 255, 261
algorithm, 207

allometric aggregation, 241
allometric aggregation approach, 194
allometric control, 266, 273, 295, 305, 307
allometric growth laws, 186
allometric relation, 182, 260, 293
allometric scaling, 243
Altemeier, 213
alveoli, 205, 244
alveoli sacs, 205
Amaral, 291
analytic functions, 98
Anderson, 158
anesthesia, 239
ante, 28
anti-correlation, 191
antipersistence, 305
antipersistent random walk, 271
antipersistent signal, 305
anxiety, 33
áperiodic behavior, 96
apprentice, 7
archery, 81
aristocracy, 9, 22
Aristotle, 90
Arnold, 106
arrow, 81
ART, 200, 221
artist, 132
ATP, 205
atrium, 58
attractor, 108, 223
Auerbach, 149, 256
Auerbach's Law, 150
aura, 298
autoregulation, 298
AV node, 60
average, 54, 70, 220
average heart rate, 64
average rate, 290, 303
average respiratory frequency, 65
average temperature, 65
average value, 123, 140, 174

Babloyantz, 236
Bachelier, 126
Badger, 156

baker, 106
balance regulation, 223
Barabási, 248, 251, 261
barrier, 241
basal ganglia, 217
Bassingthwaighte, 233
beat-to-beat intervals, 179
beating heart, 242
beetles, 149
bell-shaped curve, 77, 81, 123, 155, 173, 249
bell-shaped distribution, 102
Bernard, 38
Bernoulli, 23, 26, 100, 163, 268
Bernoulli scaling, 165, 272
Bernoulli's law, 91
Berra, 258
Beth Israel Hospital, 220
betting, 16
bifrucation path, 96
bifurcation, 96
bile duct system, 206
binary alternatives, 34
binary code, 34
binary path, 21
biological evolution, 123
biological medicine, 134
biological noise, 218, 241
biology, 147
biomechanical, 306
bit per step, 36
blood, 205
blood cells, 286
blood flow, 49, 297, 311
blood pressure, 37
blood vessels, 206
body, 49
body temperature, 37, 41, 174, 226, 242
bombs, 82
bones, 139
Boston University, 295
bowel, 206
Boyle, 74
Boyle's law, 74, 79
branch, 181
branching, 176

branching structure, 59
breath interval, 214
breath-to-breath, 209
breathing, 50, 53, 205, 242
breathing intervals, 279
breathing pattern, 239
breathing period, 56
breathing rate, 143, 221
bronchial airways, 244
bronchial tree, 205
bronchioles, 205
Brown, 269
BRV, 209, 211, 213, 220, 239, 265
BTV, 226, 265
bubble, 96
Buchman, 312

cadence, 305
Calabrese, 239
calculus of probability, 32
Cannon, 38
capillary network, 205
carbon dioxide, 41, 51
cardiac cycle, 65
cardiac dynamics, 292
cardiac pacemaker, 62
cardiopulmonary bypass pump, 311
cardiovascular control, 294
cardiovascular disease, 284
cardiovascular system, 173
casino, 28
cat's head, 107
catastrophic failures, 159
causality, 91
CBF control, 298
cell, 39, 91
central limit theorem, 161
central pattern generators, 307
cerebellum, 217
cerebral blood flow (CBF), 297
cerebral hemodynamics, 299
certainty, 91
cervix, 43
chain of command, 307
Chance, 1
chaos, 71, 72, 93, 106, 116, 297

chaotic dynamics, 283
chaotic heart, 101
characteristic scale, 206
Chardack, 62
Cheyne-Stokes syndrome, 285
child birth, 43
Chin Dynasty, 289
chronic disease, 292
church, 9, 33
cities, 149
classical scaling, 208
clinical trajectory, 313
cluster, 55, 167, 209
clustering, 220
clustering coefficient, 253
coin, 14
coin toss, 14, 21
Collins, 223
colored noise, 198, 202
complex, 70, 80, 120
complex dynamics, 290
complex hierarchial systems, 44
complex network, 243, 247, 255
complex phenomena, 3, 44, 70, 92, 93, 98, 113, 121, 151, 163, 242, 250, 302
complex system, 3, 40, 120
complex world, 69
complexity, 3, 70, 91, 112, 134, 216
complexity barrier, 241
complexity curve, 113
complexity of language, 153
computational complexity, 113
congestive heart failure, 288
Constantanople, 22
Constantinople, 23
control, 240
control mechanisms, 41, 50
control parameter, 96
control process, 264
control variable, 41
Copenhagen interpretation, 126
coronary arteries, 58
correlation, 191
cortical neurons, 302
counting, 20
creativity, 138

critical exponent, 257
critical number, 248
critical point, 103
curiosity, 33
cybernetics, 264, 302
cyclic behavior, 250

d'Alambert, 268
d'Alembert, 25, 26, 30
da Vinci, 216, 244
Darwin, 83
Daston, 23
Dawes, 209
De Lucca, 223
de Moivre, 81, 138
de Morgan, 180
de Movire, 13
deafferentiation, 309
death, 135, 224, 274, 289
decision, 42
Descartes, 95
Destexhe, 236
determinism, 72
deterministic, 89
deterministic randomness, 102
diastolic, 37
Diderot, 25
diffusion, 277
disease, 23, 135, 244, 283
disorder, 115
distribution function, 80
doctrine of equivalence, 9
don Juan, 156
donor heart, 297
Doob, 163
Doppler ultrasonograph, 299
Duke Medical Center, 211
dynamic equations, 101
dynamical equations, 120, 221
dynamical interactions, 71
dynamical variability, 221
dynamical variable, 200
dynamics, 90

ECG, 63, 295
ECG signal, 265

economics, 123, 147
EEG, 236
efficiency, 172
Einstein, 126, 270
elderly, 217
electrocardiograph, 239
electrogastrography (EGG), 229
energy, 47, 66
ensemble, 79
entropy, 34, 115, 216
epilepsy, 287
epileptic seizures, 97
equal opportunity, 33
equally likely, 36
equilibrium, 47, 278
Erdös, 248, 252
Erdös number, 252
erratic vibration, 71
error, 77, 199, 208, 241, 271
error tolerance, 208
Esen, 294
ethics, 130
Euclid, 45
evaporation, 47, 206
evolution, 39, 154
evolutionary advantage, 258
exercise, 239
exotic statistics, 255
expectation value, 21
expected gain, 27
expected winings, 27
experiment, 218
external driver, 308
eye, 38

fairness, 8, 9, 16, 33
feedback, 214, 245, 302
feedback control, 302
feedforward, 214
Fenster, 22
fetal breathing, 209
fever, 37
Feynman, 125
flowering plants, 149
fluctuations, 79, 93, 123, 129, 173, 207, 255
foot, 217

Index • 331

force, 75
forecast, 200
formation, 305
fractal, 100, 106, 207
fractal breathing, 204
fractal control, 281
fractal curve, 110
fractal dimension, 176, 191, 219, 290, 300
fractal gait, 216
fractal growth, 207
fractal gut, 228
fractal heartbeats, 191
fractal measure, 284
fractal model, 207, 244
fractal network, 99
fractal neurons, 231
fractal physiology, 72, 238, 262, 283
fractal processes, 206
fractal properties, 149
fractal respiration, 101
fractal rigidity, 301
fractal scaling, 208, 293, 311
fractal statistical phenomena, 221
fractal statistics, 1, 243
fractal structure, 108
fractal temperature, 223
fractal time series, 283
fractal variability, 307
fractional Brownian motion, 175, 199, 270, 280
fractional Lévy motion, 175
Frankenstein, 130
Franklin, 154, 251
Freeman, 97
future, 33

gait, 101, 302
gait cycle, 216, 308
gait data, 302
gait interval, 37
gait regulation, 221
Galileo, 73, 97
Galton, 83
Galvani, 231
gambler, 11, 33, 69
gambling, 9

game, 27
game of dice, 10
games of chance, 9
gaming, 23
gasping, 96
gastric rate variability (GRV), 229
Gatan Dugas, 158
Gaton, 257
Gattlletly, 239
Gauss, 77, 108, 138, 155, 173, 274
Gauss' curve, 85
genes, 139
gentleman, 11
genus, 147
geometrical fractal, 179, 181
Gernstein, 234
Gibbs, 251
Gilot, 132
glands, 264
global stability, 171
Gödel, 45
gold piece, 27
Goldberger, 177, 204, 220, 275, 284, 289, 295
Granovetter, 252
gravity, 75
great pox, 22
Greatbatch, 61
Griffin, 217
GRV, 265
GRV time series, 231

Hälsingland farm, 4
Halley, 30
Hamlet, 127
Harvard Medical School, 220
Harvard University, 295
Hausdorff, 217, 220, 309
health, 66, 177, 283
healthcare program, 135
healthy aging, 240
heart block, 61
heart rate, 37, 61, 143, 221
heart transplant, 297
heartbeat, 161
heat, 47

heavy tails, 159
heavy-tailed distribution, 109
Helmholtz, 232
hematopoiesis, 286
hierarchial, 313
Hilbert, 45
His-Purkinje, 59
histogram, 161
HIV, 169
Hoagland, 226
holistic medicine, 135
holistic perspective, 121, 241
homeostasis, 38, 49, 283, 302
homeostatic control, 40, 97, 114, 243, 266
homeostatic feedback, 247, 264
homosexual, 158
Hooke, 51
horoscopes, 9
hospital, 70
HRV, 159, 202, 209, 220, 239, 242, 265, 284, 291
human heart, 57, 93
Huntington's disease, 309
Huxley, 186
Hyperbaric Laboratory, 210
hypothesis, 73
hysteresis, 313

illness, 135
immune system, 66
income, 109, 145
incomplete knowledge, 102
inertia, 75
infection, 66
information, 32, 42, 65, 82, 115, 122, 277
information age, 122
information compression, 207
information faster, 279
information measure, 34, 116, 277
information per step, 36
information theory, 34, 153
Ingenhousz, 270
inhibition, 38
inoculation, 23
insects, 241
insurance, 12

interbeat interval, 279, 289
interchangeability, 9
intermittency, 96, 109
intermittent chaos, 198
inverse power-law, 109, 123, 149, 173, 213, 244, 256, 269, 287
inverse power-law correlation, 281
inverse power-law distribution, 123, 151, 278, 310
inverse power-law spectrum, 283
Iron Lung, 51
irregular, 88, 102
Ivanov, 292

Johnson, 127
Joyce, 152

Kac, 125, 190
Kandel, 232
Kantelhardt, 213
kidney, 206
killed virus vaccine, 52
Klafter, 166
knowability, 115
knowledge, 71, 121
Kolmogorov, 115

La Jolla Institute, 128
laboratory, 220
Lady Mary Montagu, 22
Las Vegas, 108
Latka, 297
law, 9, 33
law of cooling, 224
law of errors, 82, 122, 138, 249
law of gases, 74
law of large numbers, 26
laws of chance, 91
Lévy, 162, 280
Lévy distribution, 165, 175
Lévy random walk, 272
Lévy-flight, 166, 185
Leibniz, 24
leukemia, 287
leukopoiesis, 287
libration, 201

life expectancy, 30
life science, 95
Liljeros, 157
limit cycle, 105
Lindberg, 292
Lindenberg, 258
links, 248, 255
little ideas, 2
liver, 224
locomotion, 223, 303
logic, 73
long-range, 121
long-range interactions, 243
long-time correlations, 209, 296
long-time memory, 266, 303
Lord Rayleigh, 81
Lord Rutherford, 95
Lorenz, 89, 103
Lorenz model, 104
Lorenz system, 103
loss of complexity, 284, 311
loss of versatility, 295
Lotka, 169, 256
Lotka's law, 278
lottery, 69
luck, 3, 69
lungs, 40, 51, 205

MacDonald, 245
macroscopic law, 92
magician, 125
maladaptive, 292
mammalian lung, 206
Mandelbrot, 152, 163, 178, 234, 269
manic depression, 287
mapmakers, 125
mapping function, 201
mass, 260
mathematical expectation, 28
mathematical fractal, 181
mathematician, 10
maximum complexity, 116
maximum disorder, 278
maximum entropy, 277, 278
maximum information, 36
May, 158

mean, 187
measure of complexity, 115
measurement error, 241
medical phenomena, 114, 121
medicine, 72, 168
metabiologic, 135
metabolic activity, 298
metabolic rate, 260
metabolic reactor, 259
metronome, 305
microscopic dynamics, 123
microscopic laws, 92, 121
microscopic vital force, 270
migraine, 297
mind, 134
misfortune, 33
model of locomotion, 307
modulating, 240
monofractals, 300
Montroll, 6, 7, 30, 156, 251
Moon, 210
morphogenesis, 96
mortality curves, 31
mortality table, 29
motocontrol process, 307
motocontrol system, 174, 245
motoneurons, 309
moving average, 202
multifractal, 290
multifractal complexity, 291
multifractal phenomenon, 293
multifractal spectra, 305
multifractal spectrum, 292
multifractality, 284
multiple organ dysfunction syndrome (MODS), 312
multiple scales, 98
multiplicative, 155
murder mystery, 223
Murray, 244
Murray's Law, 244
Murrow, 131
Mutch, 213, 311

National Institutes of Health, 220
natural fractal, 181

natural law, 73, 82, 89, 111
natural philosophers, 76
nature, 124
nausea, 298
negative entropy, 277
negative-feedback, 243, 266, 274
negentropy, 115, 277
nerve conduction velocity, 309
nerve fibers, 264
network, 243, 253
network of neurons, 308
network theory, 243
neural networks, 206
neuroautonomic blockage, 291
neurodegeneration, 274
neurodegenerative diseases, 302
neuromuscular control, 309
neurons, 232, 245
neurophysiology, 114
neuroscience, 261
neutrophil count, 287
Newton, 13, 24, 73, 125
nodes, 243
noise, 71
non-traditional truths, 72, 95
nonlinear, 71, 92, 97, 200
nonlinear dynamics, 1, 113
nonlinear equations, 102
nonlinear interactions, 93
nonlinear oscillator, 60
nonlinear system, 122, 201
nonlinear world, 89
normal sinus rhythm, 60, 63, 93, 217, 266
normal walking, 303
nurse, 37

oasis, 122
odds, 11
orbit, 200
order, 33, 120
organ, 40
organization, 117
Oscar II, 88
oscillatory dynamics, 286

oxygen, 41, 51
oxytocin, 43

pacemaker, 229
pacemaker cells, 60, 93
parasympathetic control, 293
Pareto, 108, 149, 256
Pareto Principle, 259, 262
Pareto's Law, 150
Pareto's law of income, 145
Parkinson's disease, 287
path, 15, 36
pathological periodicities, 285, 287
pathological phenomena, 284
pathology, 54, 142
pattern, 40, 189
payoff, 16
Pearson, 81
pendulum, 201
Peng, 197, 213, 220, 271
periodic, 97
periodic pathology, 286
persistence, 304
phase space, 108, 201
phase transitions, 121
phase-space plot, 215
phenomena, 14
physical laws, 258
physical phenomena, 80
physical therapist, 37
physician, 37, 244
physics, 45
physiologic complexity, 283
physiologic systems, 39
physiology, 176, 206
PhysioNet, 220, 294, 303
placenta, 206
poetry, 127
Poincaré, 86, 106, 126
Poisson, 184, 249
polio, 52, 133
popularity, 257
Porter, 156
postural feedback, 264
postural sway, 223

potato, 7
power curve, 183
power-law index, 149, 175, 260
power-law relation, 183
power-law scaling, 293
predictability, 82
predictable, 175
prediction, 80, 89, 101
probabilitic reasoning, 34
probability, 20, 141
probability theory, 21, 25
proprioception, 309
proprioceptive, 217
psychological reality, 32
psychology, 100
psychology of risk, 32
psychophysics, 100

QRS complexes, 63
QRST, 193
qualitative, 96
qualitative science, 122
quantitative science, 122
quantum, 123
Quetelet, 82, 156

radiation, 47
Rall, 244
Rall's Law, 244
ramified network, 99
random, 56, 97, 243
random failures, 258
random flight, 167
random fluctuations, 174, 303
random fractal, 101
random network, 248
random noise, 40
random numbers, 55
random perturbations, 258
random phenomena, 113
random process, 213
random universe, 249
random variability, 216
random variable, 85
random walk, 36, 81, 166, 268, 277

randomness, 70, 72, 289
rank-ordering, 149
rate of breathing, 55
Rayleigh distribution, 81
reality, 16, 124
reductionism, 89, 121
reductionistic theory, 122
reflexes, 309
regularity, 71
regulation, 245
relative frequency, 20, 26, 159
REM sleep, 213
Rényi, 248
respiration, 204
respiratory controller, 214
respiratory frequency, 54
respiratory muscle, 51
respiratory period, 239
respiratory system, 143, 174
retina, 38
revolution, 9
rhetoric, 130
rhythmic output, 266
rich get richer, 256
Richardson, 169
Rigney, 204
Rössler, 107
robustness, 208
Rockefeller University, 128
Roman, 31
Rossitti, 299
rotation, 201
RR-interval time series, 194

SA node, 60
Sabine vaccine, 52
Salk, 52, 130, 168
Salk Institute, 52, 135
Salk vaccine, 52
Saltzman, 103
scale-free breathing, 263
scale-free complexity, 206
scale-free distribution, 208
scale-free Lévy, 272
scale-free nature, 256

scale-free network, 243, 260
scaling, 98, 179, 207, 223, 240, 305
scaling age, 122
scaling behavior, 277, 297
scaling exponent, 183, 290
scaling index, 276, 277
scaling principles, 110
scaling relation, 245
Schrödinger, 115
science, 69, 111
scientific breakthrough, 126
scientific complexity, 310
scientific hypothesis, 73
scientific theory, 73, 126
SCPG, 307
self-organization, 121
self-similar, 110
self-similarity, 206, 209
separatrix, 201
set point, 264
sexual liaisons, 157
Shannon, 34, 115, 152
Shlesinger, 166
shuffle, 55, 65, 108, 214
signal-plus-noise, 71
simple system, 80
sinus arrhythmia, 239
sinus node, 191
six degrees of separation, 253
skeletal muscles, 217
Slaughter's Café, 10
sleep, 213
sleep apnea, 239
sleeping, 50
small world, 253
small-world model, 255
smallpox, 2, 23, 25, 159
smokers, 284, 293
social equality, 2
sociology, 95, 123
Socrates, 127
solar system, 88
speciation, 147
spectra, 275
spectral measure, 277
spectral reserve, 276, 285

spectrum, 213, 269, 293
spinal control, 307
spinal cord, 214
SRV, 218, 220, 265, 279, 285, 302, 307
St. Petersburg game, 164
stability, 33
stance phase, 217
standard deviation, 55, 140, 187
standing, 293
Stanley, 295
state space, 199
statistical fluctuations, 50
statistical fractal, 167, 182, 241
statistical laws, 93
statistics, 20, 77
Stephensen, 299
steps, 217
stethoscope, 265
stochastic, 89
stomach, 228
strange attractor, 106
stress, 303
stress mechanism, 307
stride interval, 217, 302, 279
stride rate, 221
stride rate variability, 218, 302
Strogatz, 252
Sugihara, 204
Suki, 197
super central pattern generator, 307
supine, 293
surrogates, 297
survivor benefits, 31
swarming, 241
Swift, 180
sympathetic blockage, 292
sympathetic control, 292
system-of-systems, 112, 263, 311
systems theory, 121, 242
systolic, 37
Szeto, 209

Taylor, 182, 187
Tehrani, 53
Teich, 235
temperature, 37, 48, 224

Tennessee Williams, 127
theory of medicine, 1
theory of probability, 28
therapeutic technique, 51
therapist, 54
thermodynamics, 115, 250
thermoregulation, 49
Thom, 96
Thompson, 96
three-body problem, 89
tidal volume, 53
time intervals, 65
time of death, 224
time series, 1, 54, 67, 104, 161, 181, 217, 277
traditional truths, 72, 94
trajectories, 18
trajectory, 26, 80, 104, 214, 268
transient, 103
treadmill, 37
trees, 245
tug-of-war, 271
turbulence, 114
Turkey, 22

Ulysses, 152
uncertain, 79
uncertainty, 92
uncorrelated random process, 296
unfair process, 35
universality, 220
unpredictable, 88, 122
unstable, 92, 102
urban growth, 123, 155

vaccine, 169
van der Mark, 60

van der Pol, 60
variability, 67, 70, 142, 173, 179, 217
variable frequency, 308
variance, 220
variation, 50
ventilation, 211, 311
ventilator, 50, 51
ventricle, 58
ventricular fibrillation, 288
Vesalius, 51
Vierordt, 216
Voltaire, 23

wager, 14, 32
walking, 217, 223, 242, 266, 283, 302
Wang Shu-He, 177
Watts, 250, 255
weakly reductionistic, 124
Weaver, 28
Webber, 214
Weiss, 252
West, 204, 210, 217, 240, 258, 270, 300
white blood cells, 287
white water, 241
Wiener, 44, 115, 236, 264
Willis, 149, 186, 256
Willis' Law, 150
Woiwod, 187
World Wide Web (WWW), 247

Yamamoto, 239

Zhang, 299
Zipf, 152, 256
Zipf's Law, 152, 278
Zumofen, 166